HOW TO DESIGN AND BUILD HOME SAUNAS

A DIY Guide to Constructing Your Own Relaxation Oasis

THE FIX-IT GUY

Copyright © 2024 by The Fix-It Guy

All rights reserved. No part of this book may be reproduced in any form or by any electronic or mechanical means, including information storage and retrieval systems, without permission in writing from the publisher, except by a reviewer who may quote brief passages in a review.

Table of Contents

Introduction

Chapter 1: Introduction to Home Saunas
- History and Cultural Significance
- Health Benefits of Regular Sauna Use
- Types of Saunas: Dry, Steam, and Infrared

Chapter 2: Planning Your Home Sauna
- Assessing Space Requirements
- Choosing the Right Type for Your Needs
- Budgeting and Cost Considerations
- Obtaining Necessary Permits and Approvals

Chapter 3: Designing Your Sauna
- Layout and Ergonomics
- Ventilation and Insulation
- Lighting and Ambiance
- Materials Selection for Different Sauna Types

Chapter 4: Building a Traditional Dry Sauna
- Foundation and Framing
- Installing Insulation and Vapor Barriers
- Interior Finishing and Benches
- Heater Installation and Safety Considerations

Chapter 5: Constructing a Steam Sauna
- Waterproofing Techniques
- Drainage Systems
- Steam Generator Installation
- Tile Work and Finishing

Chapter 6: Creating an Infrared Sauna
- Understanding Infrared Technology
- Placing Infrared Panels
- Wiring and Electrical Considerations
- Customizing for Optimal Performance

Chapter 7: Sauna Accessories and Enhancements
- Seating Options and Ergonomics
- Thermometers and Hygrometers
- Aromatherapy and Essential Oils
- Audio Systems and Entertainment

Chapter 8: Maintenance and Care
- Cleaning Procedures for Different Sauna Types
- Wood Treatment and Preservation
- Troubleshooting Common Issues
- Extending the Lifespan of Your Sauna

Chapter 9: Sauna Safety and Best Practices
- Temperature and Duration Guidelines
- Hydration and Health Precautions
- Fire Safety and Electrical Considerations
- Child and Pet Safety Measures

Chapter 10: Enhancing Your Sauna Experience
- Traditional Sauna Rituals
- Combining Hot and Cold Therapies
- Meditation and Relaxation Techniques
- Sauna Socializing and Etiquette

Conclusion

Appendices:
A. Sauna Wood Types and Their Properties
B. Heater and Stove Comparison Guide
C. Glossary of Sauna Terms
D. Resources for Sauna Enthusiasts

Introduction

Have you ever stepped into a sauna and felt the immediate sense of relaxation wash over you, only to wish you could have that experience in your own home? Perhaps you've considered building a personal sauna but felt overwhelmed by the technical aspects and variety of options available. I've been there, and I understand both the allure of a home sauna and the hesitation to embark on what seems like a complex project.

Like many home improvement enthusiasts and wellness seekers, I once viewed home saunas as either basic prefab units or expensive, professionally-installed luxuries. The prospect of designing a space, choosing between dry, steam, or infrared types, understanding heating systems, and integrating it all within existing structures seemed daunting. But then I discovered the empowering world of DIY sauna building, and it changed everything.

Imagine being able to create your own personal oasis of relaxation and health, tailored exactly to your preferences and space constraints. Picture yourself confidently designing the layout, installing the perfect heating system, and customizing every detail to suit your needs. With "How to Design and Build Home Saunas," you'll gain the knowledge and skills to do just that.

This comprehensive guide is designed to equip you with expert techniques, insider tips, and step-by-step instructions for creating home saunas from start to finish. Whether you're a homeowner looking to add a touch of luxury to your daily routine or a contractor aiming to expand your service offerings, this book will take you from novice to sauna expert.

Here's what you'll discover within these pages:
- Essential tools and materials for successful sauna projects, and how to use them effectively
- Techniques for space planning, insulation, and ventilation for optimal sauna performance
- Secrets to achieving professional-level results in dry, steam, and infrared sauna construction
- Step-by-step guidance for installing different types of sauna heaters and control systems
- Expert advice on integrating saunas with existing home structures and utilities
- Troubleshooting tips for common sauna construction and operation challenges
- Professional strategies for enhancing the sauna experience with lighting, aromatherapy, and accessories

By investing in this guide, you're not just buying a book – you're unlocking the potential to create your own personal wellness retreat, increase your property value, and perhaps even start a new career. You'll save money on expensive professional installations and gain the satisfaction of building a custom sauna with your own hands.

So why settle for occasional spa visits or limited prefab options when you can become a home sauna expert yourself? Dive in now, and discover how to turn any space into a rejuvenating sanctuary, expand your DIY skills, and perhaps even turn your new expertise into a profitable business venture.

Are you ready to revolutionize your approach to home wellness and create a sauna that will be the envy of friends and family alike? Let's embark on this exciting journey together and unlock the secrets of professional-quality home sauna design and construction.

Chapter 1
Introduction to Home Saunas
History and Cultural Significance

Have you ever wondered about the origins of the sauna, or why this simple yet profound practice has endured for thousands of years? Perhaps you've experienced the rejuvenating effects of a sauna session and want to understand its deeper cultural roots. Don't worry if you're not familiar with the rich history behind this wellness tradition – many people enjoy saunas without realizing the fascinating journey that brought them into our modern lives.

The sauna is more than just a hot room; it's a testament to human ingenuity in harnessing heat for health and social bonding. At its core, the sauna represents a universal human desire for purification, relaxation, and community. It's a practice that has evolved from primitive sweat lodges to sophisticated home wellness centers, always maintaining its essence of heat-induced transformation.

But what exactly makes the sauna such a significant cultural phenomenon? Let's break it down:

1. Ancient Origins: The concept of the sweat bath dates back thousands of years, with evidence of use in various forms across many ancient civilizations. From Native American sweat lodges to Roman baths, humans have long recognized the benefits of induced sweating.

2. Finnish Innovation: While many cultures had sweat bathing traditions, it was the Finns who perfected and popularized the sauna as we know it today.

The word "sauna" itself is Finnish, reflecting the central role it plays in their culture.

3. Social Bonding: In many cultures, especially in Finland, the sauna has traditionally been a place for social gathering and important discussions. It's where families bond, friends catch up, and even business deals are sometimes struck.

4. Spiritual Significance: Many cultures view the sauna as a place of purification and rebirth. The intense heat and subsequent cooling are often seen as a metaphor for life's challenges and renewal.

5. Health and Wellness: Throughout history, saunas have been associated with physical and mental health benefits, from improved circulation to stress relief. This association has only strengthened with modern scientific research.

The cultural significance of saunas extends far beyond their physical structure:

- In Finland, saunas are considered an essential part of life, with an estimated 3.3 million saunas for a population of 5.5 million.
- During times of war, Finnish soldiers would build makeshift saunas even in the harshest conditions, demonstrating its importance to their well-being.
- In many Nordic countries, major life events like births and weddings often involve sauna rituals.
- The sauna has become a symbol of Finnish national identity and is often one of the first things built when settling in a new place.

Moreover, the sauna tradition isn't just about physical structure. It encompasses a whole set of customs and etiquette:

- The practice of alternating between hot and cold, often involving dips in cold water or snow
- The use of birch branches (vihta or vasta) to gently whip the skin, stimulating circulation
- The tradition of sauna nudity in many cultures, symbolizing equality and leaving societal roles at the door

As you embark on your journey to build a home sauna, remember that you're not just constructing a room – you're connecting with a rich tapestry of human history and wellness tradition. Whether you're a homeowner looking to add a touch of luxury to your daily routine or a wellness enthusiast aiming to deepen your understanding of heat therapy, appreciating the history and cultural significance of saunas adds depth to your experience.

In the following chapters, we'll dive deeper into the different types of saunas, design considerations, and construction techniques. You'll learn how to create a space that not only provides physical benefits but also captures the essence of this time-honored tradition. So, are you ready to bring thousands of years of wellness wisdom into your own home? Let's begin this fascinating journey into the world of saunas!

Health Benefits of Regular Sauna Use

Have you ever wondered why people emerge from saunas looking refreshed and invigorated? Or perhaps you've heard claims about saunas improving health but weren't sure of the specifics? Let's dive into the fascinating world of sauna health benefits, backed by scientific research and centuries of traditional wisdom.

Cardiovascular Health:
Regular sauna use can significantly benefit your heart and blood vessels. Studies have shown that frequent sauna bathing (4-7 times per week) is associated with:
- Reduced risk of cardiovascular disease
- Lower blood pressure
- Improved circulation

How it works: The heat exposure causes your heart rate to increase, similar to moderate exercise. This can strengthen your heart over time.

Tip: If you have a pre-existing heart condition, consult your doctor before starting a sauna routine. Start with shorter, less intense sessions and gradually build up.

Detoxification:
Sweating in a sauna helps your body eliminate toxins through the skin.
- Heavy metals like lead, zinc, and nickel can be excreted through sweat
- May help clear out environmental pollutants

How it works: The intense heat causes profuse sweating, which carries toxins out of your body.

Tip: Always shower after a sauna session to wash away the toxins on your skin. Drink plenty of water before and after to replenish fluids lost through sweating.

Stress Reduction and Mental Health:
Sauna use can have profound effects on your mental well-being:
- Releases endorphins, your body's natural feel-good chemicals
- Can reduce levels of cortisol, the stress hormone
- May help alleviate symptoms of depression and anxiety

How it works: The heat and quiet environment promote relaxation, while the physiological changes induced by the sauna can positively affect mood.

Tip: Practice mindfulness or meditation during your sauna session to enhance the stress-reducing benefits.

Skin Health:
Regular sauna use can contribute to healthier, clearer skin:
- Increased blood flow brings more nutrients to the skin
- Sweating can help clear pores and reduce acne
- May improve skin elasticity and reduce signs of aging

How it works: The heat causes blood vessels in the skin to dilate, increasing circulation and nutrient delivery.

Tip: Gently exfoliate your skin before a sauna session to maximize the cleansing benefits. Be sure to moisturize afterwards to prevent dryness.

Pain Relief and Muscle Recovery:
Athletes and chronic pain sufferers alike can benefit from sauna use:
- Can reduce muscle soreness after exercise
- May help alleviate symptoms of arthritis and fibromyalgia
- Increases production of heat shock proteins, which help repair damaged muscle fibers

How it works: Heat penetrates deep into muscles and joints, promoting blood flow and reducing inflammation.

Tip: For muscle recovery, try alternating between the sauna and a cool shower to enhance the anti-inflammatory effect.

Respiratory Health:
Sauna use may improve lung function and provide relief from respiratory conditions:
- Can help clear mucus from the lungs
- May reduce symptoms of asthma and bronchitis
- Potentially strengthens the immune system against respiratory infections

How it works: The heat and steam (in wet saunas) can help open airways and loosen mucus.

Tip: If you have a respiratory condition, start with brief sessions and pay close attention to how you feel. Some people find dry saunas more comfortable for breathing.

Improved Sleep:
Regular sauna sessions can contribute to better sleep quality:
- The drop in body temperature after a sauna can promote drowsiness
- Stress reduction effects can help calm a busy mind before bed

How it works: The sauna mimics your body's natural temperature changes before sleep, potentially helping to regulate your sleep-wake cycle.

Tip: Try using the sauna in the evening, a few hours before bedtime, for optimal sleep benefits.

Cognitive Function:
Emerging research suggests sauna use may have long-term cognitive benefits:
- Associated with reduced risk of dementia and Alzheimer's disease
- May improve focus and mental clarity

How it works: The increased blood flow to the brain and the production of brain-derived neurotrophic factor (BDNF) may contribute to these effects.

Tip: Combine regular sauna use with other brain-healthy habits like exercise and a balanced diet for maximum cognitive benefits.

Troubleshooting Tips:
1. Dehydration: If you feel dizzy or nauseous, you may be dehydrated. Exit the sauna immediately and drink water. In future sessions, hydrate well before and after.

2. Overheating: If you feel uncomfortably hot or your heart is racing excessively, leave the sauna and cool down gradually. Next time, start with a shorter session or lower temperature.

3. Skin irritation: If you experience itching or rashes, try a cooler, shorter session. Use fragrance-free soap when showering after the sauna to avoid further irritation.

4. Breathing difficulties: If you're struggling to breathe, especially in a steam sauna, try a dry sauna instead. Always listen to your body and exit if you feel uncomfortable.

5. Medication interactions: Some medications can affect how your body responds to heat. Always consult your doctor if you're on any medications before starting a sauna routine.

Remember, while saunas offer numerous health benefits, they're not suitable for everyone. Pregnant women, people with certain heart conditions, and those with some skin diseases should avoid saunas or consult a doctor first. Always listen to your body and start slowly when incorporating sauna use into your wellness routine.

By understanding these health benefits and potential challenges, you can make the most of your sauna experience, whether you're building one at home or using a public facility. Ready to unlock the power of heat for your health and well-being?

Types of Saunas: Dry, Steam, and Infrared

Have you ever wondered about the differences between the various types of saunas you've encountered? Perhaps you've experienced one type and are curious about how the others compare. Let's explore the world of saunas, breaking down the three main types - dry, steam, and infrared - to help you understand their unique characteristics, benefits, and potential drawbacks.

1. Dry Saunas (Finnish Saunas)

The traditional Finnish sauna is what most people think of when they hear the word "sauna."

How it works:
- Uses a wood or electric stove to heat rocks
- Temperatures typically range from 160°F to 200°F (71°C to 93°C)
- Humidity is low, usually 10-20%
- Users can pour water over the hot rocks to create bursts of steam (löyly)

Benefits:
- Intense heat promotes profuse sweating
- Can reach higher temperatures than other sauna types
- Authentic traditional experience
- Versatile - can be used dry or with added steam

Potential drawbacks:
- High heat can be uncomfortable for some users
- May be too intense for those with certain health conditions

Tip: Start with shorter sessions at lower temperatures and gradually work your way up to hotter, longer sessions as your body acclimates.

2. Steam Saunas (Turkish Baths or Russian Banyas)

Steam saunas, also known as wet saunas, offer a different experience from their dry counterparts.

How it works:
- Uses a steam generator to fill the room with hot, moist air
- Temperatures are lower, typically 100°F to 120°F (38°C to 49°C)
- Humidity is very high, often 100%

Benefits:
- Moist heat can be more comfortable for some users
- Excellent for respiratory health
- Can help clear skin pores more effectively
- Lower temperatures may be safer for those sensitive to extreme heat

Potential drawbacks:
- High humidity can make it feel hotter than it is
- May be uncomfortable for those with breathing difficulties
- Higher risk of bacterial growth if not properly maintained

Tip: Bring a small towel to wipe away excess sweat, and consider wearing flip-flops to avoid slipping on wet surfaces.

3. Infrared Saunas

The newest type of sauna, using a different heating mechanism altogether.

How it works:
- Uses infrared heaters to emit radiant heat
- This heat is absorbed directly by the body rather than heating the air
- Temperatures are lower, typically 120°F to 140°F (49°C to 60°C)
- Humidity is the same as the surrounding environment

Benefits:
- Can induce sweating at lower temperatures
- May penetrate deeper into tissues
- Often more energy-efficient than traditional saunas
- Can be easier to install in homes due to lower power requirements

Potential drawbacks:
- Doesn't provide the traditional sauna experience
- Limited research on long-term effects compared to traditional saunas
- Some users miss the ritual of creating steam

Tip: Experiment with the positioning of the infrared panels to find the most comfortable and effective arrangement for your body.

Comparison of Sauna Types:

1. Heat Intensity:
Dry > Steam > Infrared

2. Humidity:
Steam > Dry > Infrared

3. Energy Efficiency:
Infrared > Steam > Dry

4. Traditional Experience:
Dry > Steam > Infrared

5. Ease of Home Installation:
Infrared > Dry > Steam

6. Maintenance Requirements:
Steam > Dry > Infrared

Choosing the Right Sauna for You:

Consider these factors when deciding which type of sauna is best for your needs:

1. Personal preference: Some people prefer the intense dry heat of a Finnish sauna, while others find the moist heat of a steam room more comfortable.

2. Health considerations: If you have respiratory issues, a steam sauna might be beneficial. For those sensitive to high heat, an infrared sauna could be a better option.

3. Space and installation requirements: Infrared saunas are often easier to install in homes, while steam saunas require more complex plumbing and ventilation systems.

4. Energy efficiency: If you're concerned about energy consumption, an infrared sauna might be the most economical choice.

5. Intended use: For athletic recovery, infrared saunas are popular. For a traditional social experience, a Finnish sauna might be preferable.

Troubleshooting Tips:

1. Dry Sauna:
- If the heat feels too intense, sit on a lower bench where it's cooler.
- To increase humidity, pour small amounts of water on the rocks more frequently rather than a large amount at once.

2. Steam Sauna:
- If you're having trouble breathing, sit near the door where the air is less humid.
- To prevent slips, use textured floor tiles and provide handrails.

3. Infrared Sauna:
- If you're not sweating as expected, check the placement of the infrared panels and ensure you're within their effective range.
- If you experience any skin irritation, increase your distance from the infrared emitters.

General Tips:
- Always start with shorter sessions (10-15 minutes) and gradually increase as your body adapts.
- Stay hydrated by drinking water before, during (if comfortable), and after your sauna session.
- If you feel dizzy, nauseous, or uncomfortable, exit the sauna immediately.
- Allow your body to cool down naturally after a session before taking a cold shower.

Remember, regardless of the type you choose, regular sauna use can provide numerous health benefits. The "best" sauna is ultimately the one that you enjoy using consistently and safely. Whether you're building a home sauna or choosing which type to use at a spa, understanding these differences will help you make an informed decision and get the most out of your sauna experience. Ready to sweat your way to better health?

Chapter 2
Planning Your Home Sauna
Assessing Space Requirements

Embarking on the journey of building your own home sauna is exciting, but it all starts with careful planning. One of the most crucial steps is assessing your space requirements. Let's dive into this process with a detailed, step-by-step guide to help you make informed decisions about your sauna's size and location.

Step 1: Determine the Number of Users

Before you start measuring spaces, consider how many people will typically use the sauna at once.
- For 1-2 people: A compact sauna of 3x4 feet to 4x5 feet may suffice.
- For 3-4 people: Consider a medium-sized sauna of 5x7 feet to 6x8 feet.
- For 5-6 people or more: You'll need a larger sauna, potentially 8x10 feet or bigger.

Tip: Always plan for a bit more space than you think you'll need. It's better to have extra room than to feel cramped.

Step 2: Choose a Location

Consider these potential locations for your home sauna:
- Basement: Often ideal due to existing water-resistant flooring and proximity to plumbing.
- Garage: Can be converted relatively easily but may require additional insulation.
- Bathroom: Convenient for plumbing but may limit size options.

- Outdoor: Offers more flexibility in size but requires weatherproofing.

For each potential location, consider:
- Proximity to water supply and drainage
- Electrical access
- Ventilation possibilities
- Privacy

Step 3: Measure Your Available Space

Once you've identified potential locations:
1. Measure the length, width, and height of the area.
2. Note any obstacles like windows, doors, or fixtures.
3. Consider the space needed for changing and cooling off outside the sauna.

Remember: You'll need a minimum ceiling height of 7 feet inside the sauna, with 8 feet being ideal.

Step 4: Plan for Proper Clearances

Ensure you have adequate space around your sauna for:
- Door swing (typically outward for safety)
- Ventilation gaps (usually 4-6 inches from walls and ceiling)
- Heater clearances (check manufacturer specifications)
- Bench depth (typically 2 feet for comfort)

Step 5: Consider Auxiliary Spaces

Don't forget to plan for:
- Changing area: Ideally, allocate at least 3x4 feet for this.
- Shower space: If not already present nearby.
- Relaxation area: A cool-down space enhances the sauna experience.

Step 6: Draft a Preliminary Layout

Using graph paper or digital design software:
1. Sketch out your sauna space to scale.
2. Include benches, heater location, and door placement.
3. Ensure there's comfortable space between the top bench and ceiling (usually 42-48 inches).

Step 7: Consider Future Flexibility

Think about:
- Potential for expanding the sauna later
- Dual-purpose spaces (e.g., a sauna that can convert to a guest room)
- Resale value implications of your chosen location and size

Step 8: Consult Building Codes

Before finalizing your plans:
1. Check local building codes for any restrictions on sauna installations.
2. Verify if you need permits for electrical or plumbing work.
3. Ensure your plan complies with any homeowners' association rules.

Step 9: Get Professional Input

If you're unsure about any aspects:
1. Consult a contractor or sauna specialist.
2. Have an electrician assess your electrical capacity.
3. Consider a plumber's input if you're installing a steam sauna.

Troubleshooting Tips:

1. Limited Space:
   - Consider a corner sauna design to maximize space efficiency.
   - Look into modular or prefab sauna kits designed for small spaces.
   - Explore infrared saunas, which often have smaller footprints.

2. Low Ceilings:
   - Opt for a horizontal heater placement instead of vertical.
   - Consider a "L" shaped bench configuration to maximize seating with lower height requirements.

3. Moisture Concerns:
   - If placing the sauna in a moisture-sensitive area, plan for a vapor barrier and proper ventilation.
   - Consider a freestanding sauna unit that can be sealed against moisture intrusion.

4. Inadequate Electrical Capacity:
   - Consult an electrician about upgrading your electrical panel.
   - Consider a lower-powered infrared sauna if upgrading isn't feasible.

5. Difficult Ventilation:
   - Explore mechanical ventilation options if natural ventilation is challenging.
   - Consider a ceiling fan or small exhaust fan to improve air circulation.

6. Tight Door Clearance:
- Look into sliding door options instead of swinging doors.
- Consider an outward-opening door with special hinges that allow it to open fully against a wall.

Remember, proper planning at this stage will save you time, money, and headaches down the line. Take your time, be thorough, and don't hesitate to seek expert advice when needed. Your perfect home sauna is within reach – it just takes careful consideration and smart space assessment.

Are you ready to move forward with your sauna plans? The next steps will involve choosing the right type of sauna for your space and needs, which we'll cover in the following sections. Let's turn your sauna dreams into reality!

Choosing the Right Type for Your Needs

Selecting the perfect sauna type for your home is a crucial decision that will impact your enjoyment and the benefits you receive. Let's walk through a comprehensive process to help you make an informed choice that aligns with your needs, preferences, and space constraints.

Step 1: Understand Your Options

Refresh your knowledge of the three main sauna types:
1. Traditional Dry Sauna (Finnish)
2. Steam Sauna (Turkish Bath/Russian Banya)
3. Infrared Sauna

Step 2: Assess Your Primary Goals

Consider why you want a sauna:
- Relaxation and stress relief
- Physical recovery and muscle relaxation
- Detoxification
- Cardiovascular health
- Skin health
- Social gatherings

Different sauna types may excel in certain areas. For example:
- For muscle recovery: Infrared saunas are popular among athletes
- For respiratory health: Steam saunas can be beneficial
- For traditional experience: Finnish dry saunas are ideal

Step 3: Evaluate Your Space

Consider your available space and its characteristics:
- Size limitations
- Ventilation capabilities
- Moisture resistance of surrounding areas
- Electrical capacity
- Plumbing accessibility (especially for steam saunas)

Step 4: Consider Installation Complexity

Rank the sauna types based on installation ease in your space:
1. Infrared: Often easiest, with pre-fab options available
2. Dry Sauna: Moderate complexity
3. Steam Sauna: Most complex due to moisture management needs

Step 5: Assess Your Budget

Factor in both initial and long-term costs:
- Installation costs (highest for steam, lowest for infrared)
- Energy consumption (infrared typically most efficient)
- Maintenance requirements (steam saunas need most upkeep)

Step 6: Think About Usage Patterns

Consider how you'll use the sauna:
- Frequency of use
- Typical duration of sessions
- Number of users at one time
- Desire for quick heat-up (infrared is fastest)

Step 7: Health Considerations

Consult with a healthcare provider if you have:
- Cardiovascular conditions
- Respiratory issues
- Skin sensitivities
- Chronic health conditions

They can advise on which sauna type might be most beneficial or if any should be avoided.

Step 8: Test Different Types

Before making a final decision:
1. Visit spas or gyms with different sauna types
2. Experience each type multiple times
3. Pay attention to how you feel during and after each session

Step 9: Create a Comparison Chart

Make a table listing pros and cons of each type based on your research and experiences. Include factors like:
- Heat intensity and type
- Humidity levels
- Health benefits
- Installation requirements
- Operating costs
- Maintenance needs

Step 10: Make Your Decision

Based on your chart and personal preferences, choose the sauna type that best aligns with your needs, space, and budget.

Detailed Considerations for Each Type:

1. Traditional Dry Sauna:
Pros:
- Authentic sauna experience
- High heat for intense sweating
- Option to add steam (löyly)

Cons:
- Higher energy consumption
- Longer heat-up time
- May be too intense for some users

2. Steam Sauna:
Pros:
- Beneficial for respiratory health
- Gentler heat, suitable for longer sessions
- Great for skin hydration

Cons:
- Complex installation with moisture management
- Higher maintenance needs
- Potential for bacterial growth if not properly cared for

3. Infrared Sauna:
Pros:
- Energy-efficient
- Quick heat-up time
- Lower temperatures, more comfortable for some

Cons:
- Different experience from traditional saunas
- Limited social aspect due to typically smaller sizes
- Directional heat may feel uneven

Troubleshooting Tips:

1. Limited Space:
 - Consider a compact infrared sauna or a custom-built unit that fits your specific area
 - Look into convertible sauna designs that can be used for other purposes when not in sauna mode

2. Tight Budget:
 - Start with a portable infrared sauna and upgrade later
 - Consider a DIY kit for a traditional sauna to save on installation costs

3. High Energy Costs:
 - Opt for an infrared sauna or invest in a well-insulated traditional sauna
 - Install a timer to automatically shut off the sauna after a set duration

4. Moisture Concerns:
 - If you're worried about humidity affecting your home, lean towards a dry or infrared sauna
 - For steam saunas, ensure proper ventilation and moisture barriers are installed

5. Sensitivity to Heat:
 - Start with an infrared sauna, which operates at lower temperatures
 - In traditional saunas, sit on the lower bench where it's cooler

6. Difficulty Deciding:
 - Consider a hybrid sauna that combines traditional and infrared heating elements

- Start with a portable or rental unit to test your preferences before committing to a permanent installation

7. Multiple Users with Different Preferences:
 - Look into dual-heater saunas that can operate in both traditional and infrared modes
 - Consider a larger traditional sauna with varying height benches to accommodate different heat preferences

Remember, the "right" sauna is the one that you'll use regularly and enjoy. Take your time with this decision, as it's a significant investment in your health and home. Don't hesitate to consult with sauna specialists or health professionals if you need additional guidance.

Are you leaning towards a particular sauna type? The next steps will involve detailed planning and preparation for your chosen sauna style, which we'll cover in upcoming sections. Let's continue on your path to creating the perfect home sauna experience!

Budgeting and Cost Considerations

Creating a realistic budget for your home sauna project is crucial for its success. Let's break down the process into manageable steps and explore various cost factors to help you plan effectively.

Step 1: Determine Your Overall Budget

Start by deciding how much you're willing to invest in your sauna project. Consider:
- Your disposable income
- Home improvement savings
- Potential financing options

Tip: As a general rule, allocate 10-20% more than your initial estimate for unexpected costs.

Step 2: Break Down Costs by Category

Divide your budget into these main categories:

1. Sauna Unit/Materials:
   - Pre-fabricated kit or custom build materials
   - Heater and controls
   - Benches and interior finishes
   - Door and windows

2. Installation:
   - Labor costs (if not DIY)
   - Electrical work
   - Plumbing (for steam saunas)

3. Site Preparation:
- Flooring modifications
- Wall/ceiling alterations
- Insulation

4. Accessories:
- Lighting
- Ventilation system
- Thermometer/hygrometer
- Buckets, ladles, headrests

5. Permits and Inspections:
- Building permits
- Electrical inspections

6. Ongoing Costs:
- Energy consumption
- Maintenance supplies

Step 3: Research Costs for Your Chosen Sauna Type

Average cost ranges (as of 2024, prices may vary):

1. Infrared Sauna:
- Pre-fab kit: $1,200 - $6,000
- Custom built: $3,000 - $10,000+

2. Traditional Dry Sauna:
- Pre-fab kit: $2,000 - $7,000
- Custom built: $4,000 - $15,000+

3. Steam Sauna:
- Pre-fab kit: $2,500 - $8,000
- Custom built: $5,000 - $20,000+

Note: Custom builds offer more flexibility but generally cost more than pre-fab kits.

Step 4: Get Multiple Quotes

For each aspect of the project:
1. Obtain at least three quotes from reputable suppliers/contractors
2. Ensure quotes are detailed and include all necessary components
3. Ask about warranties and after-sale support

Step 5: Consider Long-Term Costs

Factor in ongoing expenses:
1. Energy costs: Estimate monthly usage and cost
2. Maintenance: Regular cleaning supplies, wood treatment (for traditional saunas)
3. Potential repairs: Set aside a small fund for future maintenance

Step 6: Explore Cost-Saving Options

Consider these money-saving strategies:
1. DIY installation (if you have the skills)
2. Opt for a smaller sauna size
3. Choose a less expensive wood type for interior
4. Look for end-of-season sales on sauna equipment
5. Consider a portable sauna as a lower-cost entry point

Step 7: Create a Detailed Budget Spreadsheet

Use a spreadsheet to track:
1. Estimated costs for each category
2. Actual costs as you make purchases

3. Any variances between estimated and actual costs

Step 8: Plan for Contingencies

Set aside 10-20% of your total budget for unexpected expenses, such as:
1. Unforeseen structural issues
2. Upgrades to electrical systems
3. Additional waterproofing needs

Step 9: Consider Financing Options

If needed, explore:
1. Home equity loans
2. Personal loans
3. Credit cards with 0% introductory APR for large purchases
4. Manufacturer financing (for pre-fab units)

Step 10: Assess Return on Investment (ROI)

Consider how a sauna might impact your home's value:
1. Research local real estate trends
2. Consult with a realtor about potential value addition
3. Factor in personal enjoyment and health benefits

Troubleshooting Tips:

1. Budget Overruns:
   - Prioritize essential elements and postpone luxury additions
   - Look for areas where you can substitute with less expensive materials without compromising quality
   - Consider a phased approach, adding features over time

2. Unexpected Structural Issues:
- Always have a contingency fund
- Get a thorough inspection of the area before starting the project
- Be prepared to adjust your plans if major issues are discovered

3. High Energy Costs:
- Invest in superior insulation to reduce long-term energy consumption
- Consider a more energy-efficient sauna type (e.g., infrared)
- Install a timer to limit accidental extended use

4. Costly Custom Features:
- Explore pre-fab options that might offer similar features at a lower cost
- Look into modular systems that allow for future upgrades

5. Limited Budget:
- Start with a basic model and plan for future upgrades
- Consider a portable sauna as an interim solution
- Explore secondhand markets for gently used sauna equipment

6. Financing Challenges:
- Improve your credit score before applying for loans
- Consider a personal loan from credit unions, which often offer better rates
- Look into peer-to-peer lending platforms

7. Contractor Cost Variations:
- Ensure you're comparing apples to apples in quotes (same materials, scope, etc.)

- Ask for references and check previous work before choosing the lowest bid
- Be wary of quotes that are significantly lower than others – they may not include all necessary work

8. Permit-Related Costs:
- Check with your local building department early in the process to understand all required permits
- Factor in the time cost of obtaining permits, which can delay your project

Remember, a well-planned budget is key to a successful sauna project. Take the time to research thoroughly and be realistic about costs. It's better to overestimate slightly and be pleasantly surprised than to underestimate and face financial stress mid-project.

By following these steps and keeping these considerations in mind, you'll be well-prepared to manage the financial aspects of your home sauna project. Are you ready to start crunching the numbers and turning your sauna dreams into a tangible plan?

Obtaining Necessary Permits and Approvals

Navigating the world of permits and approvals can seem daunting, but it's a crucial step in ensuring your home sauna project is safe, legal, and built to last. Let's break down this process into manageable steps and provide you with the information you need to proceed confidently.

Step 1: Research Local Regulations

Begin by investigating your local building codes and zoning laws:
1. Visit your city or county's building department website
2. Look for specific regulations related to saunas or similar structures
3. Note any special requirements for electrical, plumbing, or ventilation work

Tip: Some areas may classify saunas as "accessory structures" or "home additions," so be prepared to search under various terms.

Step 2: Determine Required Permits

Common permits for sauna projects include:
1. Building permit
2. Electrical permit
3. Plumbing permit (for steam saunas)
4. Mechanical permit (for ventilation systems)

Note: Requirements vary by location and project scope. Some areas may bundle these into a single permit for small projects.

Step 3: Gather Necessary Documentation

Prepare the following for your permit application:
1. Detailed plans or blueprints of your sauna project
2. Site plan showing the sauna's location on your property
3. Specifications of the sauna heater and other major components
4. Proof of property ownership or landlord approval (if renting)
5. Contractor licenses and insurance information (if not DIY)

Step 4: Visit Your Local Building Department

Schedule a visit to discuss your project:
1. Bring all prepared documentation
2. Ask about specific requirements for sauna installations
3. Inquire about any potential zoning issues
4. Discuss timeline for permit approval

Tip: Some departments offer pre-application consultations. Take advantage of these to catch any issues early.

Step 5: Submit Your Permit Application

Follow the department's submission process:
1. Complete all required forms
2. Pay necessary fees (keep receipts for your records)
3. Submit all supporting documentation

Note: Many departments now offer online submission options.

Step 6: Await Approval and Address Feedback

The review process typically involves:
1. Initial review by department staff
2. Possible requests for additional information or changes
3. Final approval or denial

Be prepared to:
1. Respond promptly to any requests for clarification
2. Make adjustments to your plans if required
3. Resubmit updated documentation if necessary

Step 7: Display Permits and Schedule Inspections

Once approved:
1. Display permits as required (usually visible from the street)
2. Schedule necessary inspections at key project milestones
3. Keep all documentation on-site during construction

Step 8: Final Inspection and Certificate of Occupancy

After completion:
1. Schedule a final inspection
2. Address any last-minute requirements
3. Obtain a certificate of occupancy or final approval

Troubleshooting Tips:

1. Permit Denial:
   - Request a detailed explanation of why the permit was denied
   - Ask about potential modifications that would make the project approvable

- Consider hiring a permit expediter or consultant familiar with local codes

2. Unexpected Requirements:
- Be flexible and willing to adjust your plans
- Research alternative solutions that meet code requirements
- Consider professional help if requirements seem overly complex

3. Delays in Approval Process:
- Follow up regularly but politely with the building department
- Ask if there's any additional information you can provide to speed up the process
- Consider scheduling an in-person meeting to discuss any hold-ups

4. Neighbor Objections:
- Communicate openly with neighbors about your plans
- Consider adjustments that address their concerns (e.g., privacy screens)
- Be prepared to attend and present at zoning board meetings if necessary

5. Outdated or Unclear Local Codes:
- Request clarification in writing from the building department
- Propose using standards from national building codes if local codes are ambiguous
- Consider involving a local architect or engineer to help interpret requirements

6. DIY Limitations:
- Some areas restrict certain work (especially electrical) to licensed professionals

- Be prepared to hire contractors for portions of the work if required
- Consider taking relevant certification courses if allowed by local regulations

7. Historic District or HOA Restrictions:
   - Obtain written approval from your HOA or historic preservation board before applying for permits
   - Be prepared for additional design requirements or restrictions
   - Consider less visible locations for your sauna to ease approval

8. Environmental Concerns:
   - Be aware of regulations regarding tree removal or land disturbance
   - Consider having a soil test done if there are concerns about drainage or stability
   - Prepare a plan for managing construction waste and runoff

9. Permit Expiration:
   - Note the expiration date on your permits and plan your project timeline accordingly
   - If delays occur, apply for extensions before permits expire
   - Keep records of ongoing work to demonstrate project progress if needed

10. Inspection Failures:
    - Carefully review inspector comments and required corrections
    - Address all issues promptly and completely
    - Don't hesitate to ask inspectors for clarification on requirements

Remember, while obtaining permits can seem like a hassle, it's a crucial step in ensuring your sauna is safe, legal, and built to last. Proper permits can also protect you from potential legal issues and make selling your home easier in the future.

By following these steps and being prepared for potential challenges, you'll navigate the permit process more smoothly. Are you ready to take the next step in making your home sauna project official and above-board?

Chapter 3
Designing Your Sauna
Layout and Ergonomics

Designing the perfect sauna layout is crucial for creating a comfortable, functional, and enjoyable space. Let's dive into the process of planning your sauna's layout with a focus on ergonomics to ensure the best possible experience.

Step 1: Determine Sauna Size

Consider the following factors:
1. Number of users (typically allow 2 feet of bench space per person)
2. Available space in your home
3. Ceiling height (minimum 7 feet, ideally 8 feet or more)

Example sizes:
- 2-person sauna: 4' x 5' to 5' x 7'
- 4-person sauna: 6' x 8' to 8' x 8'
- 6-person sauna: 8' x 10' to 10' x 12'

Step 2: Plan Bench Configuration

Common configurations include:
1. L-shaped: Maximizes seating in smaller spaces
2. U-shaped: Ideal for larger saunas and social gatherings
3. Straight: Simple and effective for narrow spaces

Ergonomic considerations:
- Bench depth: 20-24 inches for comfort
- Bench height: 18-20 inches for the lower bench, 36-40 inches for the upper bench
- Backrest angle: 100-110 degrees for optimal relaxation

Step 3: Locate the Heater

Placement is crucial for even heat distribution:
1. Traditional heater: Usually placed near the door, about 6-12 inches off the floor
2. Infrared panels: Typically mounted on walls and/or ceiling

Safety tip: Ensure proper clearances around the heater as specified by the manufacturer.

Step 4: Design the Door

Consider these factors:
1. Direction: Outward-swinging for safety
2. Size: Typically 24 inches wide, 72-80 inches tall
3. Material: Wood or tempered glass
4. Location: Usually opposite the heater for best heat circulation

Accessibility tip: If designing for users with mobility issues, consider a wider door (30-36 inches).

Step 5: Plan Ventilation

Proper airflow is essential:
1. Inlet vent: Near the heater, about 6 inches off the floor
2. Outlet vent: Opposite wall from the heater, about 6 inches from the ceiling
3. Consider a small, quiet exhaust fan for improved air circulation

Step 6: Incorporate Lighting

Plan for ambient and task lighting:
1. Recessed ceiling lights for overall illumination
2. Under-bench lighting for a relaxing atmosphere
3. Consider color-changing LED options for chromotherapy

Safety tip: Ensure all lighting fixtures are rated for high-temperature, high-humidity environments.

Step 7: Add Accessories

Enhance functionality with:
1. Towel hooks near the door
2. A water bucket and ladle stand near the heater
3. A thermometer/hygrometer at eye level when seated
4. Headrests and back supports for added comfort

Step 8: Create a Relaxation Area

If space allows, design a cool-down area outside the sauna:
1. Include seating (chairs or benches)
2. Add a small table for water glasses
3. Consider a shower for rinsing off

Step 9: Draft Your Design

Use graph paper or design software to create a scaled drawing of your sauna, including:
1. Wall dimensions
2. Bench layout
3. Heater location
4. Door swing
5. Vent placements
6. Lighting positions

7. Accessory locations

Step 10: Review and Refine

Analyze your design for:
1. Flow: Ensure easy movement in and out of the sauna
2. Comfort: Check that seating is ample and ergonomic
3. Safety: Verify proper clearances around the heater
4. Functionality: Ensure all necessary elements are included and logically placed

Troubleshooting Tips:

1. Limited Space:
   - Consider a corner sauna design to maximize space efficiency
   - Use fold-down benches for flexibility
   - Explore vertical space with taller, narrower designs

2. Low Ceiling:
   - Opt for a horizontal heater placement
   - Use a single-level bench design
   - Consider infrared panels on walls instead of a traditional heater

3. Uncomfortable Seating:
   - Adjust bench depths or heights for better ergonomics
   - Add removable cushions or backrests for personalized comfort
   - Install adjustable-height benches for versatility

4. Poor Heat Distribution:
   - Reassess heater placement and size
   - Add heat deflectors or stones to improve heat flow

- Consider additional vents or a small fan for better circulation

5. Claustrophobia Concerns:
- Incorporate a glass door or window for openness
- Use lighter wood tones to create a sense of space
- Add indirect lighting to create depth

6. Accessibility Issues:
- Widen the door and create a no-step entry
- Install grab bars near benches and the door
- Consider a infrared sauna with a lower operating temperature

7. Inadequate Ventilation:
- Add an additional vent or increase vent sizes
- Install a small, quiet exhaust fan
- Ensure the door has a small gap at the bottom for air intake

8. Lack of Privacy:
- Position the door away from high-traffic areas
- Add frosted film to glass doors or windows
- Install a changing area or screen outside the sauna

9. Insufficient Lighting:
- Layer lighting with a combination of ambient and task lights
- Use dimmable fixtures for adjustable ambiance
- Consider fiber optic lighting for a starry-sky effect on the ceiling

10. Limited Relaxation Area:
- Use space-saving furniture like folding chairs or wall-mounted seats
- Create a multi-functional space that serves as both changing and relaxation area
- Utilize outdoor space for a cool-down area if possible

Remember, the key to a great sauna design is balancing functionality, comfort, and aesthetics. Take your time with this phase, as a well-thought-out layout will enhance your sauna experience for years to come.

Are you excited to start sketching out your sauna design? Remember, it's often helpful to create a few different layout options before settling on your final design. Which aspects of the layout are you most looking forward to customizing for your perfect sauna experience?

Ventilation and Insulation

Proper ventilation and insulation are crucial for creating a safe, efficient, and enjoyable sauna experience. Let's dive into these essential aspects of sauna design with a detailed, step-by-step guide.

Ventilation:

Step 1: Understand the Importance of Ventilation

Proper airflow in your sauna:
- Regulates temperature and humidity
- Removes excess moisture and odors
- Provides fresh air for comfortable breathing
- Helps maintain the longevity of your sauna materials

Step 2: Plan Your Ventilation System

A basic sauna ventilation system consists of:
1. Supply vent (air intake)
2. Exhaust vent (air outlet)
3. Adjustable vents or fans for air circulation

Step 3: Position the Supply Vent

Place the supply vent:
- Near the floor, close to the heater
- Size: typically 4-6 inches in diameter
- Material: use a metal vent cover rated for high temperatures

Tip: Some heater models have built-in air intakes, which can serve as your supply vent.

Step 4: Locate the Exhaust Vent

Position the exhaust vent:
- High on the wall opposite the heater, or on the ceiling
- At least 6 inches below the ceiling
- Size: typically 6-10 inches in diameter
- Material: use a moisture-resistant, heat-tolerant vent cover

Step 5: Consider Additional Circulation

For larger saunas or improved air movement:
- Install a small, heat-resistant fan near the ceiling
- Use adjustable vents to fine-tune airflow
- Consider a ventilation gap under the door (about 3/4 inch)

Step 6: Install a Vapor Barrier

Place a vapor barrier behind the insulation on walls and ceiling:
- Use 6-mil polyethylene sheeting
- Overlap seams by at least 6 inches and tape securely
- Ensure a tight seal around vents and other penetrations

Insulation:

Step 7: Choose the Right Insulation

Select insulation with these properties:
- High R-value (at least R-13 for walls, R-26 for ceiling)
- Moisture-resistant
- Fire-retardant

Options include:
- Fiberglass batts (most common)

- Mineral wool (excellent sound insulation)
- Rigid foam board (high R-value per inch)

Step 8: Insulate the Walls

Install insulation in this order:
1. Install vapor barrier on the warm side (sauna interior)
2. Place insulation between wall studs
3. Ensure a snug fit without compression
4. Use unfaced insulation to prevent double vapor barriers

Tip: Leave a small air gap (about 1/2 inch) between the insulation and the exterior wall to prevent moisture buildup.

Step 9: Insulate the Ceiling

Follow the same process as the walls, but:
- Use thicker insulation (R-26 or higher)
- Consider adding a radiant barrier for extra heat reflection

Step 10: Insulate the Floor (if applicable)

For saunas not on a concrete slab:
1. Install a vapor barrier on the subfloor
2. Add rigid foam insulation
3. Top with a layer of plywood before installing the final flooring

Step 11: Seal All Penetrations

Use high-temperature caulk or specialized flashing to seal:
- Around vent openings
- Electrical and plumbing penetrations
- Any gaps in the vapor barrier

Step 12: Test Your System

Before finishing the interior:
1. Run your sauna heater
2. Check for proper air circulation
3. Monitor humidity levels
4. Look for any signs of moisture accumulation

Troubleshooting Tips:

1. Poor Air Circulation:
   - Increase the size of vents
   - Add a small, heat-resistant fan
   - Ensure the supply and exhaust vents are not obstructed

2. Excessive Humidity:
   - Check the vapor barrier for gaps or tears
   - Increase exhaust ventilation
   - Consider adding a dehumidifier in the changing area

3. Cold Spots:
   - Check for gaps in insulation
   - Use a thermal camera to identify areas of heat loss
   - Add additional insulation to problem areas

4. Condensation on Windows or Doors:
   - Improve ventilation around these areas
   - Consider double-pane windows or adding a storm door
   - Use a squeegee to remove excess moisture after each use

5. Mold or Mildew Growth:
   - Inspect vapor barrier and insulation for moisture intrusion
   - Improve ventilation to reduce humidity
   - Use mold-resistant materials in problem areas

6. Drafts:
- Check door and window seals
- Ensure vents are properly sized (not too large)
- Add weatherstripping to reduce air leakage

7. Inefficient Heating:
- Verify that insulation R-values meet or exceed recommendations
- Check for thermal bridging through studs or other structures
- Consider adding a radiant barrier in the ceiling

8. Odors:
- Increase ventilation rates
- Ensure proper cleaning and drying after each use
- Use natural odor absorbers like charcoal or essential oils

9. Noisy Ventilation:
- Use sound-dampening insulation around vents
- Choose quiet, high-quality fans
- Ensure ductwork is properly sized to reduce air velocity

10. Difficulty Regulating Temperature:
- Install adjustable vents for better control
- Consider a programmable control system for your heater
- Adjust insulation levels in specific areas as needed

Remember, proper ventilation and insulation are key to a comfortable, efficient, and long-lasting sauna. Take the time to plan and execute these elements carefully, as they form the foundation of a great sauna experience.

Are you feeling confident about tackling the ventilation and insulation for your sauna project? Remember, it's often worth consulting with a professional or experienced sauna builder if you're unsure about any aspects of the process. What part of the ventilation and insulation process are you most interested in customizing for your sauna?

Lighting and Ambiance

Creating the right atmosphere in your sauna is crucial for relaxation and enjoyment. Let's explore how to design the perfect lighting and ambiance for your home sauna.

Step 1: Understand Lighting Goals

Consider the purposes of sauna lighting:
- Safety: Ensuring visibility for entering, exiting, and moving around
- Functionality: Allowing for reading or other activities
- Atmosphere: Creating a relaxing, spa-like environment
- Therapy: Potentially incorporating chromotherapy (color therapy)

Step 2: Choose Light Fixtures

Select fixtures designed for high-heat, high-humidity environments:
1. Recessed ceiling lights (IP65 rated or higher)
2. Wall sconces (heat-resistant)
3. LED strip lighting (silicone-encased for moisture protection)
4. Fiber optic lighting systems (heat-free light source outside the sauna)

Tip: Always choose fixtures explicitly rated for sauna use to ensure safety and longevity.

Step 3: Plan Lighting Placement

Consider these locations for optimal lighting:
1. Ceiling: For overall illumination

2. Under benches: For subtle, ambient lighting
3. Behind backrests: For a soft, glowing effect
4. Around the heater: To highlight this focal point (maintain safe distances)

Step 4: Incorporate Dimming Capabilities

Install dimmer switches or smart lighting systems to:
- Adjust light levels for different times of day
- Create various moods for different sauna experiences
- Accommodate personal preferences of multiple users

Step 5: Consider Color Temperature

Choose bulbs or LED systems with appropriate color temperatures:
- Warm white (2700K-3000K): For a cozy, relaxing atmosphere
- Cool white (3100K-4500K): For a clean, refreshing feel
- Adjustable color temperature: For versatility

Step 6: Explore Chromotherapy Options

If interested in color therapy, consider:
- LED systems with color-changing capabilities
- Dedicated chromotherapy lights
- Colored lens filters for standard fixtures

Step 7: Plan Natural Light (If Applicable)

If including windows or skylights:
- Position for privacy
- Use frosted or textured glass for diffused light
- Consider adding curtains or blinds for adjustability

Step 8: Design Control Systems

Plan for convenient light control:
- Place switches outside the sauna for pre-heating illumination
- Consider motion sensors for automatic activation
- Explore smart home integration for app-controlled lighting

Step 9: Enhance Ambiance with Materials

Choose materials that complement your lighting:
- Light-colored woods for a brighter feel
- Dark woods for a cozier atmosphere
- Textured surfaces for interesting shadow play

Step 10: Add Finishing Touches

Incorporate additional elements for perfect ambiance:
- Himalayan salt lamps for a warm glow and potential health benefits
- Battery-operated candles for a flickering effect without fire risk
- Essential oil diffuser for aromatic ambiance (place outside sauna)

Troubleshooting Tips:

1. Harsh Lighting:
   - Install dimmers for adjustable brightness
   - Use frosted bulb covers or diffusers
   - Add indirect lighting sources (e.g., under-bench LEDs)

2. Inadequate Illumination:
- Increase the number of light fixtures
- Use higher lumen output bulbs (while maintaining safe wattage)
- Add task lighting near seating areas

3. Glare on Reflective Surfaces:
- Adjust fixture angles to reduce direct reflection
- Use indirect lighting techniques
- Apply anti-glare films to problematic surfaces

4. Inconsistent Light Distribution:
- Rearrange fixture placement for more even coverage
- Use multiple smaller lights instead of few large ones
- Incorporate reflective surfaces to bounce light

5. Overheating Light Fixtures:
- Ensure all fixtures are rated for sauna use
- Create ventilation space around recessed lights
- Switch to lower heat-emitting options (e.g., LEDs)

6. Moisture Damage to Fixtures:
- Verify proper IP ratings for all components
- Improve overall sauna ventilation
- Apply additional sealant around fixture installations

7. Difficulty with Controls:
- Relocate switches for easier access
- Install smart lighting systems for remote control
- Use glow-in-the-dark switch covers for visibility

8. Color Rendering Issues:
- Choose bulbs with high CRI (Color Rendering Index) ratings
- Test different color temperatures to find the most flattering option
- Use natural light when possible for true color representation

9. Energy Efficiency Concerns:
- Switch to LED bulbs for lower power consumption
- Install occupancy sensors to avoid lights being left on
- Use timers to automatically shut off lights after typical session lengths

10. Uninspiring Atmosphere:
- Experiment with colored lighting or color-changing systems
- Add natural elements like plants (in the changing area) or stone accents
- Incorporate audio systems for soothing sounds or music

Remember, lighting and ambiance are crucial for creating the perfect sauna experience. Take time to experiment with different options to find what works best for you. Don't be afraid to adjust and refine your lighting setup over time as you discover your preferences.

Are you excited about designing the lighting for your sauna? Which aspects of lighting and ambiance are you most looking forward to personalizing in your sauna space? Remember, the goal is to create an environment that promotes relaxation and rejuvenation, tailored to your unique tastes and needs.

Materials Selection for Different Sauna Types

Choosing the right materials for your sauna is crucial for durability, safety, and the overall sauna experience. Let's explore the material selection process for different sauna types, focusing on traditional dry saunas, steam saunas, and infrared saunas.

Step 1: Understand Sauna Requirements

Before selecting materials, consider the unique needs of each sauna type:
- Dry Saunas: High heat resistance, low moisture absorption
- Steam Saunas: Excellent moisture resistance, mold resistance
- Infrared Saunas: Heat reflectivity, comfort at lower temperatures

Step 2: Choose Interior Wood

For all sauna types, select wood that is:
- Low in resin content
- Resistant to warping and cracking
- Comfortable to touch at high temperatures

Popular options include:
1. Western Red Cedar: Aromatic, naturally resistant to decay
2. Hemlock: Light color, minimal scent, affordable
3. Nordic White Spruce: Durable, light color, minimal knots
4. Basswood: Soft, light-colored, virtually knot-free

Note: Avoid woods like pine or cedar with high resin content, as they can emit unpleasant odors when heated.

Step 3: Select Benches and Interior Structures

Choose woods that are:
- Splinter-resistant
- Comfortable for bare skin
- Able to withstand frequent cleaning

Best options:
- Abachi: Extremely low heat conductivity, splinter-free
- Clear Western Red Cedar: Durable, naturally antimicrobial
- Basswood: Soft and comfortable, light color

Step 4: Choose Flooring Materials

For dry and infrared saunas:
- Wooden duck boards or tiles made from the same wood as interior
- Textured ceramic tiles for non-slip surface

For steam saunas:
- Non-porous, non-slip tiles (porcelain or ceramic)
- Pebble tiles for a spa-like feel and excellent drainage

Step 5: Select Heater Surrounds

For traditional heaters:
- Heat-resistant stones (peridotite, olivine diabase)
- Protective wood railing around heater

For infrared saunas:
- Reflective panels behind heating elements
- Heat-resistant covers for infrared emitters

Step 6: Choose Door Materials

For all sauna types:
- Tempered glass for visibility and safety
- Wood frame matching interior wood type
- Consider all-glass doors for modern aesthetics

Step 7: Select Insulation Materials

For all sauna types:
- Fiberglass batts (unfaced) for walls and ceiling
- Reflective foil insulation for enhanced heat reflection
- Rigid foam insulation for floors (if applicable)

Step 8: Choose Vapor Barrier

For dry and infrared saunas:
- 6-mil polyethylene sheeting

For steam saunas:
- Specialized steam room membrane systems

Step 9: Select Exterior Finishing

Choose based on sauna location:
- Indoor: Match surrounding room decor
- Outdoor: Weather-resistant materials (e.g., treated lumber, fiber cement siding)

Step 10: Consider Accessories

Choose materials for:
- Headrests: Matching interior wood or soft, washable fabric
- Backrests: Matching interior wood, ergonomically designed
- Light fixtures: Heat-resistant, moisture-proof materials

Troubleshooting Tips:

1. Wood Discoloration:
- Use a UV-resistant finish on exterior wood
- Install proper ventilation to reduce moisture buildup
- Clean and treat wood regularly with sauna-specific products

2. Bench Warping:
- Ensure proper support structure underneath benches
- Use quarter-sawn lumber for increased stability
- Allow wood to acclimate to sauna environment before installation

3. Floor Drainage Issues (Steam Saunas):
- Install a proper slope towards the drain (1/4 inch per foot)
- Use smaller tiles to allow for better contouring to drain slope
- Consider a linear drain system for improved water management

4. Mold Growth:
- Improve ventilation to reduce moisture
- Use naturally mold-resistant woods like cedar
- Apply a safe, sauna-specific mold inhibitor to problem areas

5. Excessive Wood Movement:
- Allow wood to acclimate to the sauna environment before installation
- Use tongue-and-groove construction for better stability
- Ensure proper spacing between boards to allow for expansion

6. Peeling Vapor Barrier:
- Use high-temperature adhesive tapes for seams
- Ensure proper overlap at seams (6 inches minimum)
- Consider professional installation for steam room membranes

7. Heater Stone Degradation:
- Use only stones specifically rated for sauna use
- Replace stones annually or as recommended by manufacturer
- Avoid pouring large amounts of water on stones at once

8. Door Seal Failures:
- Use high-temperature silicone for door seals
- Install adjustable hinges to maintain proper alignment
- Consider magnetic seals for better performance

9. Insulation Settling:
- Use friction-fit batts to prevent settling
- Install insulation baffles in attic spaces
- Consider blown-in insulation for hard-to-reach areas

10. Exterior Finish Fading (Outdoor Saunas):
- Apply UV-resistant finishes
- Install awnings or overhangs for sun protection
- Use composite materials designed for outdoor exposure

Remember, the key to successful material selection is understanding the specific requirements of your sauna type and choosing materials that can withstand the unique conditions they'll be exposed to. Don't hesitate to consult with sauna specialists or material suppliers for advice tailored to your specific project.

Are you feeling confident about selecting materials for your sauna? Which aspect of material selection are you most excited about customizing for your perfect sauna experience? Remember, investing in quality materials upfront can save you time and money in the long run by creating a durable, enjoyable sauna that will last for years to come.

Chapter 4
Building a Traditional Dry Sauna
Foundation and Framing

Creating a solid foundation and sturdy frame is crucial for building a long-lasting, safe, and efficient traditional dry sauna. Let's walk through the process step-by-step, providing you with a detailed guide to get started on your sauna project.

Step 1: Plan Your Foundation

1. Determine the sauna location:
    - Indoor: Existing floor may suffice if it's level and can support the weight
    - Outdoor: Requires a new foundation

2. For outdoor saunas, choose a foundation type:
    - Concrete slab (most common)
    - Pier foundation
    - Gravel bed with pressure-treated lumber frame

3. Size your foundation:
    - Add 6 inches on all sides beyond your planned sauna dimensions
    - Ensure at least 4 inches thickness for concrete slabs

Step 2: Prepare the Site

1. Clear the area:
    - Remove vegetation, rocks, and debris
    - Level the ground, allowing for slight slope for drainage

2. Mark the outline:
- Use stakes and string to outline the foundation
- Verify corners are square using the 3-4-5 method

3. Excavate if necessary:
- Dig to a depth of 6 inches plus the thickness of your slab
- Add 4 inches of gravel for improved drainage

Step 3: Pour the Concrete Foundation (for outdoor saunas)

1. Build the form:
- Use 2x4 lumber to create a frame for the slab
- Secure with stakes and ensure it's level

2. Add reinforcement:
- Lay a grid of rebar or wire mesh for strength

3. Pour the concrete:
- Use a 3000 PSI mix or higher
- Smooth the surface with a trowel
- Allow proper curing time (usually 7 days)

Step 4: Frame the Floor (if not using a concrete slab)

1. Create a pressure-treated lumber frame:
- Use 2x6 or 2x8 joists, 16 inches on center
- Install a rim joist around the perimeter

2. Add floor sheathing:
- Use 3/4 inch pressure-treated plywood
- Secure with deck screws

Step 5: Frame the Walls

1. Cut your wall plates:
   - Use 2x4 lumber for top and bottom plates
   - Cut to the dimensions of your sauna design

2. Assemble wall frames:
   - Use 2x4 studs, 16 inches on center
   - Include openings for door and windows

3. Raise and secure the walls:
   - Plumb each wall with a level
   - Secure to the floor with framing nails or concrete anchors

4. Install the top plate:
   - Use double top plates for added strength
   - Overlap corners for improved stability

Step 6: Frame the Ceiling

1. Determine ceiling style:
   - Flat ceiling: Use ceiling joists
   - Sloped ceiling: Install rafters

2. Install ceiling framing:
   - Use 2x6 or 2x8 lumber, 16 inches on center
   - Ensure proper support for the expected snow load (for outdoor saunas)

3. Add collar ties:
   - Install horizontal 2x4s between rafters for added strength

Step 7: Frame for Sauna Features

1. Create bench supports:
- Install horizontal 2x4s at appropriate heights for upper and lower benches

2. Frame for heater:
- Reinforce the area where the heater will be mounted
- Ensure proper clearances as per manufacturer's specifications

3. Prepare for ventilation:
- Frame openings for intake and exhaust vents

Troubleshooting Tips:

1. Uneven Foundation:
- Use a self-leveling compound to correct minor imperfections
- For major issues, consider hiring a professional to ensure proper leveling

2. Moisture Issues in Wooden Floors:
- Install a vapor barrier beneath the floor frame
- Ensure proper ventilation beneath the structure

3. Warped Framing Lumber:
- Select lumber carefully, avoiding pieces with obvious defects
- Store lumber properly before use to prevent warping

4. Inadequate Support for Heavy Heaters:
- Double up studs where the heater will be mounted
- Consider installing a horizontal support beam for extra strength

5. Difficulty Plumbing Walls:
- Use temporary braces to hold walls plumb while securing
- Invest in a long level or laser level for accuracy

6. Sagging Ceiling:
- Install additional support beams if needed
- Consider using engineered lumber for longer spans

7. Gaps in Framing Joints:
- Use shims to fill small gaps
- For larger issues, consider re-cutting lumber for a tighter fit

8. Condensation on Outdoor Slab:
- Ensure proper slope for drainage away from the sauna
- Apply a concrete sealer to reduce moisture absorption

9. Difficulty Aligning Door Frame:
- Use shims to adjust the frame for proper alignment
- Ensure the rough opening is square before installing the door

10. Weak Corners:
- Use metal framing connectors at corners for added strength
- Ensure overlap of top plates at corners for improved stability

Remember, a well-built foundation and frame are essential for the longevity and safety of your sauna. Take your time with this phase, as it sets the stage for everything that follows. Don't hesitate to consult with a professional or experienced DIYer if you encounter challenges beyond your skill level.

Are you feeling ready to tackle the foundation and framing of your traditional dry sauna? Which aspect of this process are you most excited about or perhaps a bit nervous about? Remember, careful planning and attention to detail at this stage will pay off in the long run with a sturdy, well-built sauna that you can enjoy for years to come.

Installing Insulation and Vapor Barriers

Proper insulation and vapor barrier installation are crucial for your sauna's efficiency, longevity, and performance. Let's walk through this process step-by-step, ensuring you have all the information needed for a successful installation.

Step 1: Gather Materials and Tools

Materials:
- Insulation (fiberglass batts, mineral wool, or rigid foam)
- 6-mil polyethylene sheeting for vapor barrier
- Aluminum foil tape or specialized vapor barrier tape
- Staples and staple gun
- Utility knife
- Safety gear (gloves, mask, safety glasses)

Tools:
- Measuring tape
- Straight edge or T-square
- Marker or pencil
- Scissors or utility knife for cutting vapor barrier

Step 2: Prepare the Sauna Space

1. Ensure the framing is complete and secure
2. Clean the area of debris and dust
3. Check for any protruding nails or screws

Step 3: Install Insulation

1. Measure the spaces between studs and joists
2. Cut insulation to fit, allowing for a snug fit without compression

3. Install insulation in walls:
- Start at the bottom and work your way up
- Ensure insulation fits tightly around electrical boxes and other obstacles
- Use unfaced insulation to avoid double vapor barriers
4. Install insulation in the ceiling:
- Work from one end to the other
- Be careful not to block any planned ventilation
5. If using rigid foam for the floor, install it now:
- Cut to fit the floor area
- Tape seams with foil tape

Tip: Wear protective gear when handling insulation to avoid skin irritation and inhalation of fibers.

Step 4: Install the Vapor Barrier

1. Start with the ceiling:
- Unroll the polyethylene sheeting across the ceiling joists
- Staple every 6-8 inches along the joists
- Overlap seams by at least 6 inches and tape securely
2. Move to the walls:
- Start at a corner and work your way around the room
- Overlap the ceiling vapor barrier by at least 6 inches
- Staple the barrier to the studs every 6-8 inches
- Overlap vertical seams by 6 inches and tape securely
3. For the floor (if not on a concrete slab):
- Lay polyethylene sheeting over the entire floor
- Overlap seams and tape securely
- Extend the barrier a few inches up the walls

Step 5: Seal Penetrations and Seams

1. Cut small X-shaped slits for electrical boxes and other penetrations
2. Carefully fold the flaps and staple around the penetration
3. Seal all seams, penetrations, and staple lines with aluminum foil tape

Step 6: Install a Second Layer of Vapor Barrier (Optional but Recommended)

1. Repeat the process with a second layer of polyethylene sheeting
2. Offset the seams from the first layer for better vapor protection

Step 7: Final Checks

1. Inspect all seams and penetrations for proper sealing
2. Ensure the vapor barrier is continuous from floor to ceiling
3. Check that no insulation is visible or compressed

Troubleshooting Tips:

1. Gaps in Insulation:
 - Use smaller pieces of insulation to fill gaps
 - Consider using spray foam for hard-to-reach areas

2. Vapor Barrier Tears:
 - Patch small tears with vapor barrier tape
 - For larger tears, cut a piece of polyethylene to cover the damaged area, overlapping by at least 6 inches on all sides, and tape securely

3. Insulation Falling Out:
- Use insulation supports (wire or plastic) to hold batts in place
- Consider using friction-fit batts for a tighter hold

4. Condensation Behind Vapor Barrier:
- Ensure proper ventilation in the sauna
- Check for gaps in the vapor barrier and seal thoroughly

5. Difficulty Working Around Electrical Boxes:
- Cut insulation to fit snugly around boxes
- Use spray foam to fill small gaps around electrical penetrations

6. Sagging Ceiling Insulation:
- Use insulation supports or chicken wire to hold insulation in place
- Consider using rigid foam board for easier installation

7. Vapor Barrier Too Tight or Loose:
- Allow some slack in the vapor barrier to account for wood movement
- If too loose, remove and re-staple, ensuring it's taut but not stretched

8. Insulation Compression in Tight Spaces:
- Use thinner insulation in areas with less depth
- Consider using spray foam in very tight spaces

9. Air Pockets in Vapor Barrier:
- Smooth out the barrier as you install it
- Use a plastic smoother tool to eliminate air pockets

10. Difficulty Aligning Multiple Layers:
- Use a different color marker for each layer to keep track of seams
- Consider using a helper to manage large sheets of polyethylene

Remember, proper installation of insulation and vapor barriers is crucial for your sauna's performance and longevity. Take your time and pay attention to detail, especially around seams and penetrations. A well-insulated sauna with an effective vapor barrier will be more energy-efficient and resistant to moisture damage.

Are you feeling confident about tackling the insulation and vapor barrier installation? Which part of this process do you think will be most challenging for your specific sauna design? Remember, it's always better to take extra time during this stage to ensure everything is properly sealed and insulated, as it will save you potential headaches down the line.

Interior Finishing and Benches

The interior finishing and bench installation are crucial steps that will define the look, feel, and functionality of your sauna. Let's go through this process step-by-step to create a beautiful and comfortable sauna interior.

Step 1: Prepare the Space

1. Ensure insulation and vapor barrier are properly installed
2. Clean the area thoroughly
3. Gather materials and tools:
    - Sauna-grade wood (cedar, hemlock, or aspen)
    - Stainless steel screws
    - Wood glue
    - Saw (circular saw, miter saw)
    - Drill
    - Level
    - Measuring tape
    - Sandpaper

Step 2: Install Wall Paneling

1. Start with the ceiling:
    - Install furring strips perpendicular to joists for air circulation
    - Attach tongue-and-groove boards, starting from one corner
    - Use stainless steel nails or screws, driving them through the tongue

2. Move to the walls:
    - Begin at a corner and work your way around
    - Install vertical furring strips for air gap if desired

- Use a level to ensure boards are straight
- Cut around obstacles like vents or light fixtures

Tip: Leave a 1/4 inch gap at the floor and ceiling for wood expansion.

Step 3: Install Flooring

1. If using wood flooring:
   - Lay sleepers (2x2 treated lumber) for air circulation
   - Install tongue-and-groove flooring perpendicular to sleepers
2. If using tile:
   - Install cement board over plywood subfloor
   - Apply waterproof membrane before tiling
   - Use sauna-safe grout and sealant

Step 4: Build and Install Benches

1. Determine bench layout and heights:
   - Upper bench typically 36-40 inches high
   - Lower bench 18-20 inches high

2. Build bench frames:
   - Use 2x4 lumber for frame
   - Ensure frames are level and sturdy

3. Create bench tops:
   - Use 2x4 or 2x6 lumber for seat slats
   - Space slats 1/8 to 1/4 inch apart for air circulation
   - Round edges and sand smooth for comfort

4. Install benches:
- Secure frames to wall studs for stability
- Ensure benches are level
- Add support legs if necessary for longer benches

Step 5: Install Backrests and Headrests

1. Build backrest frames and attach to wall
2. Install backrest boards at a comfortable angle (about 110 degrees)
3. Add removable headrests for extra comfort

Step 6: Build and Install Heater Guard

1. Construct a wooden guard around the heater
2. Ensure proper clearances as per manufacturer's instructions
3. Use wider gaps between slats for better heat circulation

Step 7: Add Finishing Touches

1. Install door and window trim
2. Add corner molding if desired
3. Install sauna accessories:
- Thermometer/hygrometer
- Light fixtures
- Towel hooks
- Bucket shelf

Step 8: Sand and Finish

1. Sand all surfaces to ensure smoothness
2. Apply sauna-safe wood treatment if desired (optional, as many prefer natural wood)

Troubleshooting Tips:

1. Uneven Walls or Ceiling:
- Use shims behind furring strips to create a level surface
- Consider using a laser level for more precise alignment

2. Wood Expansion Issues:
- Ensure proper gaps at floors, ceilings, and corners
- Use spacers during installation to maintain consistent gaps

3. Bench Stability Concerns:
- Add additional support legs to long benches
- Use angle brackets for extra reinforcement

4. Difficulty Cutting Around Obstacles:
- Make paper templates for complex cuts
- Use a jigsaw for intricate cutting around vents or fixtures

5. Gaps in Tongue-and-Groove Joints:
- Ensure boards are fully seated before securing
- Use a tapping block to avoid damaging board edges

6. Bench Comfort Issues:
- Round over all edges that contact skin
- Consider adding removable bench cushions for extra comfort

7. Heater Guard Heat Distribution:
- Adjust slat spacing for optimal heat flow
- Ensure the guard doesn't impede air circulation around the heater

8. Wood Discoloration:
- Use only stainless steel fasteners to prevent staining
- Apply a UV-resistant finish if the sauna will be exposed to sunlight

9. Squeaky Benches:
- Use wood glue in addition to screws for a tighter bond
- Add rubber pads under bench legs to reduce noise

10. Uneven Flooring:
- Use a self-leveling compound before installing flooring
- For severe unevenness, consider rebuilding the subfloor

Remember, the interior finish is what you'll see and interact with every time you use your sauna, so take your time to ensure everything is done to your satisfaction. Pay special attention to comfort and safety, particularly with bench construction and heater guard installation.

Are you excited to see your sauna taking shape with the interior finishes? Which aspect of the interior are you most looking forward to customizing? Remember, while following general guidelines is important, this is also your opportunity to add personal touches that will make your sauna uniquely yours.

Heater Installation and Safety Considerations

Installing your sauna heater correctly is crucial for both performance and safety. Let's walk through the process step-by-step, ensuring you have all the information needed for a safe and effective installation.

Step 1: Choose the Right Heater

1. Determine the appropriate heater size:
   - Calculate your sauna's cubic footage (length x width x height)
   - Consult manufacturer guidelines for recommended heater size
   - Factor in any windows or poorly insulated areas

2. Select heater type:
   - Electric (most common for home saunas)
   - Wood-burning (for authentic experience, requires proper ventilation)
   - Gas (less common, requires professional installation)

Step 2: Prepare for Installation

1. Gather necessary tools and materials:
   - Screwdriver set
   - Drill and bits
   - Level
   - Wire strippers
   - Voltage tester
   - Heater mounting hardware

2. Review manufacturer's installation guide thoroughly

3. Ensure electrical requirements are met:
- Verify circuit capacity (typically 220-240V, 30-60 amp dedicated circuit)
- Confirm wire gauge is appropriate for heater's power requirements

Step 3: Position the Heater

1. Follow manufacturer's guidelines for clearances:
- Typically 4-5 inches from walls
- At least 8 inches above floor
- Adequate space from benches and door

2. Mark mounting locations on wall

3. Install any required heat shields on nearby walls

Step 4: Mount the Heater

1. Securely attach mounting bracket to wall studs
2. Lift heater into place (get help if it's heavy)
3. Secure heater to mounting bracket
4. Use a level to ensure the heater is straight

Step 5: Electrical Connection

CAUTION: If you're not experienced with electrical work, hire a licensed electrician for this step.

1. Turn off power at the main circuit breaker
2. Run appropriate gauge wire from the circuit breaker to the heater location
3. Install a disconnect box near the heater for safety

4. Connect wires according to manufacturer's wiring diagram:
- Typically, you'll connect two hot wires, a neutral, and a ground
5. Double-check all connections are tight and secure

Step 6: Install Controls

1. Mount control panel outside the sauna room
2. Connect control wiring to heater according to manufacturer's instructions
3. Install temperature sensor in sauna room as directed

Step 7: Install Rocks

1. Wash sauna rocks thoroughly before installation
2. Carefully place rocks in heater, allowing for air circulation:
- Place larger rocks at the bottom
- Fill to just above heating elements, don't overfill

Step 8: Safety Checks

1. Install heater guard rail if required
2. Ensure proper ventilation is in place
3. Install a carbon monoxide detector if using a wood or gas heater

Step 9: Test the Heater

1. Turn power back on at the circuit breaker
2. Run heater through a complete heating cycle
3. Check for any unusual smells or sounds
4. Verify temperature control is working correctly

Troubleshooting Tips:

1. Heater Not Heating:
 - Check circuit breaker hasn't tripped
 - Verify all electrical connections are secure
 - Ensure temperature setting is correct

2. Uneven Heating:
 - Rearrange rocks for better air circulation
 - Check for proper clearances around heater

3. Unusual Smells:
 - Some odor is normal on first use; run heater for an hour to burn off manufacturing residues
 - If smell persists, check for any plastic or other materials contacting the heater

4. Tripping Circuit Breaker:
 - Verify the circuit is properly sized for the heater
 - Check for any short circuits in the wiring

5. Control Panel Not Responding:
 - Check connections between control panel and heater
 - Ensure temperature sensor is correctly installed and connected

6. Rocks Cracking or Exploding:
 - Use only sauna-grade rocks
 - Avoid pouring cold water on extremely hot rocks

7. Heater Overheating:
 - Check that the temperature sensor is working and properly placed
 - Ensure rocks are not blocking air vents on the heater

8. Corrosion on Heater:
- Improve ventilation in the sauna
- Avoid using chlorinated water on rocks

9. WiFi-Connected Controls Not Working:
- Check your home's WiFi signal strength in the sauna area
- Ensure app and heater firmware are up to date

10. Delayed Heating:
- Pre-heat time is normal; most saunas take 30-60 minutes to reach temperature
- If excessively slow, check for air leaks or poor insulation in the sauna room

Remember, safety is paramount when installing a sauna heater. If you're unsure about any aspect of the installation, particularly the electrical work, don't hesitate to consult or hire a professional. A correctly installed heater will provide years of safe and enjoyable sauna experiences.

Are you feeling confident about the heater installation process? Which aspect of the installation do you think will require the most attention in your sauna setup? Remember, taking the time to install your heater correctly will ensure optimal performance and safety for your sauna sessions.

Chapter 5
Constructing a Steam Sauna
Waterproofing Techniques

Proper waterproofing is crucial when constructing a steam sauna, as it protects the structure from moisture damage and ensures longevity. Let's dive into the step-by-step process of effectively waterproofing your steam sauna.

Step 1: Prepare the Surface

1. Ensure the framing is complete and sturdy
2. Clean all surfaces thoroughly:
   - Remove dust, debris, and any protruding nails or screws
   - Repair any cracks or holes in the substrate

3. Check for levelness and make adjustments if necessary

Step 2: Install Cement Board

1. Cut cement board to fit walls and ceiling:
   - Use a score-and-snap technique or a circular saw with a cement board blade
   - Leave a 1/4 inch gap at floor and ceiling for expansion

2. Attach cement board to framing:
   - Use cement board screws every 6-8 inches
   - Ensure screws are countersunk slightly

3. Tape and mud all seams:
   - Use alkali-resistant mesh tape
   - Apply thin-set mortar over tape, smooth with a trowel
   - Allow to dry completely

Step 3: Apply Waterproof Membrane

1. Choose a suitable membrane:
   - Liquid-applied membrane (e.g., RedGard)
   - Sheet membrane (e.g., Kerdi)

2. For liquid-applied membrane:
   - Apply first coat with a roller or brush
   - Allow to dry according to manufacturer's instructions
   - Apply a second coat, ensuring full coverage
   - Pay extra attention to corners, seams, and penetrations

3. For sheet membrane:
   - Apply thin-set mortar to the surface
   - Carefully lay the membrane, smoothing out any air bubbles
   - Overlap seams by at least 2 inches
   - Use manufacturer-approved sealant for seams and corners

Step 4: Waterproof Penetrations

1. Identify all penetrations (pipes, vents, etc.)
2. Use appropriate flashing or prefabricated sleeves
3. Seal around penetrations with waterproof sealant
4. Reinforce with an extra layer of membrane

Step 5: Create Proper Slope for Drainage

1. Ensure floor has a slope of 1/4 inch per foot towards the drain
2. Use a self-leveling compound to create or adjust slope if necessary

Step 6: Install Vapor Barrier (if not using a sheet membrane)

1. Apply 6-mil polyethylene sheeting to walls and ceiling
2. Overlap seams by at least 6 inches
3. Seal all seams with waterproof tape

Step 7: Waterproof the Floor

1. Apply waterproof membrane to the floor:
 - Extend membrane up the walls by at least 6 inches
 - Ensure proper integration with the drain

2. For curbless designs, extend waterproofing beyond the steam room

Step 8: Test Waterproofing

1. Plug the drain and fill the room with a few inches of water
2. Let it stand for 24-48 hours
3. Check for any leaks or water loss

Step 9: Final Sealing

1. Apply silicone sealant to all corners and transitions
2. Use steam-room grade sealant for best results

Troubleshooting Tips:

1. Membrane Bubbling:
 - For sheet membranes, carefully cut bubble, press out air, and reseal
 - For liquid membranes, allow more drying time between coats

2. Poor Adhesion:
- Ensure surface is clean and dry before application
- Use a primer if recommended by the manufacturer

3. Cracks in Cement Board:
- Reinforce with additional mesh tape and thin-set
- Consider using a crack isolation membrane

4. Water Pooling on Floor:
- Verify and adjust slope towards the drain
- Use a self-leveling compound to correct low spots

5. Leaks Around Penetrations:
- Reapply sealant and membrane around the penetration
- Consider using a prefabricated sleeve for a better seal

6. Mold Growth:
- Improve ventilation in the steam room
- Use mold-resistant products in all stages of construction

7. Peeling Membrane:
- Ensure proper surface preparation before application
- Check compatibility of all products used

8. Failed Water Test:
- Carefully inspect for the source of the leak
- Reapply membrane and sealant to suspected areas
- Consult a professional if unable to locate the leak

9. Condensation on Ceiling:
- Ensure proper insulation above the vapor barrier
- Consider installing a sloped ceiling for better water runoff

10. Sealant Discoloration:
- Use only steam room-rated sealants
- Avoid using bleach or harsh chemicals for cleaning

Remember, thorough waterproofing is essential for the longevity and proper function of your steam sauna. Taking the time to do this step correctly will prevent costly repairs and ensure a enjoyable, trouble-free sauna experience for years to come.

Are you feeling confident about tackling the waterproofing process for your steam sauna? Which aspect of waterproofing do you think will be most critical for your specific sauna design? Remember, while this process may seem time-consuming, it's one of the most important steps in creating a durable and effective steam sauna.

Drainage Systems

A properly designed and installed drainage system is crucial for a steam sauna, ensuring water is efficiently removed to maintain a clean and safe environment. Let's walk through the process of creating an effective drainage system for your steam sauna.

Step 1: Plan Your Drainage System

1. Determine drain location:
   - Usually in the center or corner of the room
   - Consider accessibility for maintenance

2. Choose drain type:
   - Linear drain (for a modern look and efficient water removal)
   - Center drain (traditional and cost-effective)
   - Multiple drains for larger saunas

3. Calculate required drainage capacity:
   - Consider steam generator output
   - Factor in shower heads if included

Step 2: Create Proper Floor Slope

1. Establish a slope of 1/4 inch per foot towards the drain
2. For linear drains, create a single-plane slope
3. For center drains, create a four-way slope

Tools needed:
- Level
- Straight edge
- Marker or chalk line

Step 3: Install the Drain

1. Cut opening in the subfloor for the drain body
2. Install drain flange:
   - Apply plumber's putty or silicone sealant under the flange
   - Secure flange to subfloor

3. Connect drain to plumbing:
   - Ensure proper P-trap installation
   - Use appropriate fittings and pipe material (PVC or ABS)

4. Test connection for leaks before proceeding

Step 4: Apply Waterproof Membrane

1. Install cement board or appropriate substrate
2. Apply waterproof membrane over entire floor and 6 inches up walls:
   - Use liquid-applied membrane or sheet membrane
   - Ensure proper integration with drain flange

3. For linear drains, follow manufacturer's instructions for membrane application

Step 5: Install Tile or Floor Finish

1. Apply thin-set mortar and install tiles:
   - Maintain proper slope during installation
   - Use spacers for consistent grout lines

2. For linear drains, ensure tile edges align properly with drain

3. Allow proper curing time before grouting

Step 6: Apply Grout and Sealant

1. Use epoxy grout for better water resistance
2. Apply grout, ensuring full coverage in joints
3. Clean excess grout promptly
4. Once cured, apply grout sealer for added protection

Step 7: Install Drain Cover

1. For center drains, choose a decorative cover that matches your design
2. For linear drains, install the manufacturer-provided grate

Step 8: Additional Drainage Features (Optional)

1. Install a secondary emergency drain if desired
2. Consider a trench drain at the entrance for transitional drainage

Step 9: Final Testing

1. Conduct a flood test:
   - Plug the drain and fill with 2-3 inches of water
   - Mark the water level and check after 24-48 hours
   - Inspect for any leaks or water loss

Troubleshooting Tips:

1. Standing Water:
   - Check and adjust floor slope
   - Ensure drain is not clogged
   - Verify drain is installed at the lowest point

2. Slow Drainage:
 - Check for proper P-trap installation
 - Ensure adequate pipe slope (1/4 inch per foot)
 - Clear any clogs in the drain or pipes

3. Leaks Around Drain:
 - Reseal drain flange with silicone or plumber's putty
 - Check for proper integration of waterproof membrane with drain

4. Cracked Tiles:
 - Ensure proper substrate preparation and curing
 - Use crack isolation membrane in areas prone to movement

5. Mold or Mildew Growth:
 - Improve ventilation in the steam room
 - Ensure proper slope for complete water drainage
 - Use mold-resistant grout and sealers

6. Grout Deterioration:
 - Reapply grout sealer regularly
 - Consider regrouting with epoxy grout for better water resistance

7. Drain Cover Corrosion:
 - Use only corrosion-resistant materials (stainless steel or brass)
 - Clean and dry drain cover after each use

8. Odors from Drain:
 - Ensure P-trap is properly installed and always filled with water
 - Clean drain regularly to prevent buildup

9. Uneven Tile Surface:
- Use a longer level during installation to ensure consistent slope
- Consider self-leveling compounds for subfloor preparation

10. Clogged Linear Drain:
- Remove grate and clean thoroughly
- Use a drain snake if clog is deeper in the system

Remember, an effective drainage system is essential for the proper function and longevity of your steam sauna. Proper installation will prevent water damage, mold growth, and ensure a comfortable and hygienic environment.

Are you feeling prepared to tackle the drainage system installation for your steam sauna? Which aspect of the drainage system do you think will be most crucial for your specific sauna design? Remember, while this process requires attention to detail, it's a fundamental step in creating a functional and enjoyable steam sauna experience.

Steam Generator Installation

Installing a steam generator is a critical step in creating your steam sauna. This process requires careful planning and execution to ensure safe and efficient operation. Let's walk through the installation process step-by-step.

Step 1: Choose the Right Steam Generator

1. Calculate the required capacity:
    - Measure your sauna's cubic footage (length x width x height)
    - Factor in materials (tile, glass, etc.) and any windows
    - Consult manufacturer guidelines for appropriate size

2. Consider features:
    - Auto-flush systems for easier maintenance
    - Aromatherapy options
    - Digital controls

Step 2: Select the Installation Location

1. Choose a dry, well-ventilated area outside the steam room:
    - Within 25 feet of the steam room (closer is better)
    - Accessible for maintenance
    - Protected from freezing temperatures

2. Ensure adequate clearance around the unit as per manufacturer's instructions

Step 3: Prepare for Installation

Gather necessary tools and materials:
- Drill and bits
- Pipe wrench
- Teflon tape
- Copper pipes and fittings
- Electrical wiring and conduit
- Stainless steel steam head

Step 4: Install Water Supply Line

1. Connect a dedicated 3/8" copper line to cold water supply
2. Install a shut-off valve for easy maintenance
3. Use a water hammer arrester to prevent pipe damage
4. Consider installing a water softener if you have hard water

Step 5: Install Drain Line

1. Connect a 3/4" copper drain line
2. Ensure proper slope (1/4" per foot) for drainage
3. Install a vacuum breaker to prevent backflow

Step 6: Steam Line Installation

1. Use 3/4" copper pipe for the steam line
2. Slope the line towards the steam room (1/4" per foot)
3. Insulate the steam line to prevent heat loss and condensation
4. Install the steam head 6-12 inches above the floor, away from seating areas

Step 7: Electrical Connection

CAUTION: This step should be performed by a licensed electrician.

1. Install a dedicated circuit for the steam generator
2. Use appropriate gauge wire based on the unit's amperage
3. Install a disconnect switch near the unit for safety

Step 8: Control Panel Installation

1. Mount the control panel inside the steam room:
   - 4-5 feet above the floor
   - Away from the direct steam path

2. Run low-voltage wiring from the control panel to the generator

Step 9: Temperature Sensor Installation

1. Install the temperature sensor in the steam room:
   - 5-6 feet above the floor
   - Away from the direct steam path and cold air sources

2. Connect sensor wiring to the generator

Step 10: Final Connections and Testing

1. Make final plumbing and electrical connections
2. Turn on water supply and check for leaks
3. Power on the unit and test all functions
4. Allow system to run through a complete cycle

Troubleshooting Tips:

1. Generator Not Producing Steam:
   - Check water supply and electrical connections
   - Verify control panel settings
   - Ensure heating elements are functioning

2. Inadequate Steam:
   - Confirm generator size is appropriate for room volume
   - Check for steam line leaks or poor insulation
   - Verify steam head is not clogged

3. Water Leaks:
   - Tighten all plumbing connections
   - Check for cracked fittings or pipes
   - Ensure proper sealing around steam head

4. Erratic Temperature Control:
   - Verify temperature sensor placement and connection
   - Check for drafts or cold spots in the steam room
   - Calibrate the control system if necessary

5. Loud Noises:
   - Check for water hammer in supply lines
   - Ensure proper slope on steam and drain lines
   - Verify all mounting brackets are secure

6. Slow Heat-Up Time:
   - Inspect steam line for proper insulation
   - Check for adequate power supply
   - Verify room is properly insulated and sealed

7. Control Panel Not Responding:
- Check low-voltage wiring connections
- Ensure panel is not exposed to direct water spray
- Consider replacing the control panel if issues persist

8. Mineral Buildup:
- Implement regular flushing and descaling routine
- Consider installing a water softener
- Use manufacturer-recommended cleaning products

9. Steam Head Dripping:
- Ensure proper slope on steam line
- Check for condensation due to poor insulation
- Verify steam head is installed at the correct height

10. Automatic Flush Not Working:
- Check drain line for clogs or improper slope
- Verify solenoid valve operation
- Ensure adequate water pressure for flushing

Remember, while DIY installation is possible for those with advanced skills, hiring a professional for steam generator installation is often recommended due to the complexity of plumbing and electrical work involved. This ensures safety, proper function, and may be necessary to maintain your warranty.

Are you feeling prepared to oversee or assist with the steam generator installation for your sauna? Which aspect of the installation process do you think will require the most attention in your setup? Remember, a correctly installed steam generator is key to creating a luxurious and reliable steam sauna experience.

Tile Work and Finishing

Tile work is the final touch that brings your steam sauna to life, providing both functionality and aesthetic appeal. Let's walk through the process of tiling and finishing your steam sauna, ensuring a beautiful and durable result.

Step 1: Prepare the Surface

1. Ensure waterproofing is complete and fully cured
2. Clean the surface thoroughly
3. Check for levelness and make any necessary adjustments

Step 2: Plan Your Tile Layout

1. Measure the room and create a layout plan
2. Use a tile calculator to determine the number of tiles needed
3. Consider tile patterns (e.g., straight lay, herringbone, mosaic)
4. Dry-lay tiles to visualize the final look and adjust as needed

Step 3: Gather Materials and Tools

- Tiles (porcelain or ceramic recommended for steam rooms)
- Thin-set mortar
- Grout (epoxy grout is best for steam rooms)
- Tile spacers
- Notched trowel
- Tile cutter or wet saw
- Rubber float
- Grout sponge
- Bucket and mixing tools

Step 4: Mix and Apply Thin-Set

1. Mix thin-set according to manufacturer's instructions
2. Apply thin-set to the surface using a notched trowel
3. Work in small sections to prevent premature drying

Step 5: Install Tiles

1. Start from the center of the room or a focal point
2. Press tiles firmly into the thin-set
3. Use spacers to maintain consistent grout lines
4. Cut tiles as needed to fit around obstacles or edges
5. Clean excess thin-set from tile surfaces as you go

Step 6: Allow Proper Curing Time

1. Let thin-set cure for 24-48 hours (follow manufacturer's recommendations)
2. Avoid any foot traffic or moisture during this time

Step 7: Mix and Apply Grout

1. Remove spacers
2. Mix epoxy grout according to instructions
3. Apply grout using a rubber float, pressing into joints
4. Work in small sections, especially with epoxy grout

Step 8: Clean Tiles and Finish Grout

1. Use a damp sponge to remove excess grout
2. Rinse sponge frequently and change water as needed
3. Polish tiles with a microfiber cloth to remove haze

Step 9: Apply Sealant

1. Once grout is fully cured, apply a penetrating sealer to grout lines
2. Use silicone caulk in corners and where tiles meet other surfaces

Step 10: Install Finishing Touches

1. Mount shower fixtures, if applicable
2. Install glass doors or curtain rods
3. Add seating, if not built-in

Troubleshooting Tips:

1. Uneven Tile Surface:
   - Use a level during installation to ensure evenness
   - Consider using tile leveling systems for large format tiles

2. Grout Discoloration:
   - Ensure consistent mixing of grout
   - Avoid using too much water when cleaning
   - Consider using color-sealed grout for uniformity

3. Cracked Tiles:
   - Ensure proper substrate preparation
   - Use crack isolation membranes in areas prone to movement
   - Allow for expansion joints in large areas

4. Lippage (Uneven Tile Edges):
   - Use tile leveling systems
   - Choose rectified tiles for tighter grout lines
   - Take extra care with large format tiles

5. Grout Haze:
- Clean tiles thoroughly before grout sets
- Use a grout haze remover if necessary
- Polish with a microfiber cloth after cleaning

6. Poor Tile Adhesion:
- Ensure proper mixing and application of thin-set
- Check that tiles are pressed firmly into the mortar
- Verify that thin-set hasn't skinned over before placing tiles

7. Efflorescence (White Powder on Grout):
- Use a penetrating sealer on grout
- Improve ventilation in the steam room
- Clean with a mild acid solution if it occurs

8. Mold Growth in Grout:
- Use epoxy grout or antimicrobial grout additives
- Ensure proper ventilation and regular cleaning
- Apply grout sealer annually

9. Grout Cracking:
- Allow for proper curing time
- Use flexible grout in areas subject to movement
- Ensure correct grout mixing ratios

10. Tile Lipping on Walls:
- Use a straightedge or level to check alignment frequently
- Start tiling from the bottom up for walls
- Use tile spacers consistently

Remember, proper tile installation is crucial for both the appearance and functionality of your steam sauna. Take your time, work carefully, and don't hesitate to seek professional help for complex patterns or large areas.

Are you excited to see your steam sauna come to life with the tile work? Which tile design or pattern are you considering for your sauna? Remember, while tiling can be a rewarding DIY project, it requires patience and attention to detail. If you're unsure about any aspect, consulting with or hiring a professional tiler can ensure a flawless finish that will last for years to come.

Chapter 6
Creating an Infrared Sauna
Understanding Infrared Technology

Infrared saunas have gained popularity due to their unique heating method and potential health benefits. Let's dive into the technology behind infrared saunas to help you understand how they work and why they're different from traditional saunas.

What is Infrared Technology?

Infrared technology uses light waves to produce heat. This light is part of the electromagnetic spectrum and is invisible to the human eye. There are three types of infrared waves:

1. Near-infrared (NIR): Shortest wavelength, penetrates the skin surface
2. Mid-infrared (MIR): Medium wavelength, penetrates deeper into soft tissue
3. Far-infrared (FIR): Longest wavelength, penetrates deepest into the body

How Infrared Saunas Work:

1. Infrared emitters produce infrared light waves
2. These waves penetrate the body directly, heating it from within
3. The surrounding air remains relatively cool compared to traditional saunas
4. This allows for longer sessions and potentially deeper sweating at lower temperatures

Key Components of Infrared Saunas:

1. Infrared Emitters: The heat source, usually carbon or ceramic
2. Reflective Panels: Help direct infrared waves towards the user
3. Control Panel: Regulates temperature and session duration
4. Insulation: Keeps heat focused inside the sauna

Benefits of Infrared Technology in Saunas:

1. Lower Operating Temperature: Typically 120-140°F vs. 150-180°F in traditional saunas
2. Direct Body Heating: May lead to more efficient sweating
3. Deeper Heat Penetration: Potentially offering therapeutic benefits for muscles and joints
4. Energy Efficiency: Often uses less electricity than traditional electric saunas
5. Faster Heat-Up Time: Usually ready in 15-20 minutes

Considerations When Choosing Infrared Technology:

1. Full Spectrum vs. Far Infrared:
- Full spectrum includes near, mid, and far infrared
- Far infrared is most common and deeply penetrating

2. EMF Levels:
- Look for low EMF (electromagnetic field) emitters
- Some brands offer ultra-low EMF technology

3. Heat Distribution:
- Consider the placement and number of emitters for even heating

4. Material Safety:
- Ensure all materials used are heat-safe and non-toxic when heated

5. Control Options:
- Look for adjustable temperature and timer settings
- Some models offer preset programs or smart device control

Step-by-Step Guide to Understanding Your Infrared Sauna:

1. Research different infrared technologies and their benefits
2. Determine which type of infrared (NIR, MIR, FIR, or full spectrum) suits your needs
3. Understand the heating pattern of different emitter types (carbon, ceramic, etc.)
4. Learn about safe exposure times and temperatures for infrared saunas
5. Familiarize yourself with proper maintenance and cleaning procedures

Troubleshooting Tips:

1. Uneven Heating:
- Check emitter placement and functionality
- Ensure reflective panels are properly installed
- Consider adding more emitters for better coverage

2. Slow Heat-Up Time:
- Verify proper electrical connection and voltage
- Check for any obstructions blocking emitters
- Ensure the sauna is well-insulated

3. High EMF Levels:
- Measure EMF levels with a gaussmeter
- Contact manufacturer about low-EMF options
- Consider repositioning seating further from emitters

4. Malfunctioning Control Panel:
- Check all wiring connections
- Ensure firmware is up to date (for digital controls)
- Contact manufacturer for replacement if needed

5. Uncomfortable Eye or Skin Sensations:
- Adjust emitter placement to avoid direct exposure
- Use protective eyewear if necessary
- Start with shorter sessions and gradually increase duration

6. Condensation Issues:
- Improve ventilation in the sauna room
- Use a small fan to circulate air
- Wipe down surfaces after each use

7. Odors During Operation:
- Ensure all materials are sauna-grade and heat-safe
- Clean the sauna thoroughly with non-toxic cleaners
- Run empty heat cycles to "burn off" any manufacturing residues

8. Inconsistent Temperature Readings:
- Calibrate the temperature sensor
- Ensure sensor is placed correctly in the sauna
- Use a secondary thermometer to verify readings

Remember, understanding the technology behind your infrared sauna is key to maximizing its benefits and troubleshooting any issues that may arise. As you plan your infrared sauna project, consider which aspects of the technology are most important to you. Are you looking for full-spectrum heating, or is far-infrared sufficient for your needs? How important is low EMF emission in your choice of emitters?

By thoroughly understanding infrared technology, you'll be better equipped to design, build, and enjoy your personal infrared sauna. Are you feeling more confident about the science behind infrared saunas now? What aspect of infrared technology are you most excited to incorporate into your sauna design?

Placing Infrared Panels

Properly placing infrared panels is crucial for creating an effective and comfortable infrared sauna experience. Let's walk through the process of planning and installing your infrared panels step-by-step.

Step 1: Plan Panel Placement

1. Determine the number of panels needed:
    - Consider room size and desired heat intensity
      - Consult manufacturer guidelines for recommended coverage

2. Sketch a layout plan:
    - Aim for even heat distribution
    - Common placements: walls, corners, ceiling, and under benches

3. Consider user comfort:
    - Ensure panels won't be too close to seated users
    - Plan for even body coverage when seated

Step 2: Prepare the Space

1. Ensure the sauna room is properly insulated
2. Install proper electrical wiring:
    - Consult an electrician if needed
    - Ensure adequate power supply for all panels

3. Clean and clear the installation areas

Step 3: Gather Tools and Materials

- Infrared panels
- Mounting brackets (usually provided with panels)
- Screwdriver or drill
- Level
- Pencil for marking
- Stud finder
- Safety equipment (gloves, safety glasses)

Step 4: Locate Wall Studs

1. Use a stud finder to locate studs in the walls
2. Mark stud locations lightly with a pencil

Step 5: Mount Brackets

1. Hold mounting brackets against the wall at desired height
2. Use a level to ensure brackets are straight
3. Mark screw holes on the wall
4. Pre-drill holes if necessary
5. Secure brackets to studs using provided screws

Step 6: Install Wall Panels

1. Carefully lift the infrared panel
2. Align panel with mounted brackets
3. Secure panel to brackets following manufacturer instructions
4. Ensure panel is level and firmly attached

Step 7: Install Ceiling Panels (if applicable)

1. Follow similar process as wall panels

2. Use assistance when lifting and mounting overhead panels
3. Ensure extra secure mounting for safety

Step 8: Install Under-Bench Panels (if applicable)

1. Mount brackets to bench frame or floor
2. Ensure adequate airflow around the panel
3. Take care not to block any ventilation

Step 9: Connect Electrical

CAUTION: If you're not experienced with electrical work, hire a licensed electrician for this step.

1. Turn off power at the main circuit breaker
2. Connect panels to the electrical supply as per manufacturer instructions
3. Ensure all connections are secure and properly insulated

Step 10: Test the System

1. Turn on power and activate each panel
2. Check for even heating and proper function
3. Use an infrared thermometer to verify heat output

Troubleshooting Tips:

1. Uneven Heating:
 - Adjust panel placement for better coverage
 - Check for any obstructions blocking infrared rays
 - Ensure all panels are functioning correctly

2. Panels Not Heating:
 - Verify electrical connections

- Check control panel settings
- Consult manufacturer for potential panel malfunction

3. Overheating:
- Ensure proper spacing between panels and users
- Verify temperature control settings
- Check for any objects too close to panels

4. Cold Spots:
- Add additional panels to problem areas
- Adjust panel angles for better coverage
- Improve room insulation

5. Electrical Issues:
- Verify circuit capacity is adequate for all panels
- Check for loose connections
- Consider upgrading electrical supply if necessary

6. Panel Instability:
- Reinforce mounting brackets
- Ensure panels are secured to studs, not just drywall
- Double-check all mounting hardware is tight

7. Interference with Bench Seating:
- Adjust panel placement to avoid contact with users
- Consider using lower-wattage panels in tight spaces

8. Condensation on Panels:
- Improve room ventilation
- Check room insulation for cold spots
- Consider adding a small fan for air circulation

9. Inefficient Heating:
- Verify room is properly sealed and insulated
- Ensure door seals tightly when closed
- Pre-heat sauna before use for optimal performance

10. Control Panel Issues:
- Check connections between control panel and panels
- Ensure control panel is not exposed to excessive heat or moisture
- Update firmware if available

Remember, proper placement of infrared panels is key to creating a comfortable and effective sauna experience. Take your time with the planning and installation to ensure optimal performance and safety.

Are you excited about setting up your infrared sauna? Which panel configuration are you considering for your space? Remember, while DIY installation is possible, don't hesitate to consult with a professional if you're unsure about any aspect, especially when it comes to electrical work. The goal is to create a safe, comfortable, and effective infrared sauna that you can enjoy for years to come.

Wiring and Electrical Considerations

Proper electrical setup is crucial for the safe and efficient operation of your infrared sauna. While some aspects of this process may require a licensed electrician, understanding the basics can help you plan and communicate effectively. Let's walk through the key considerations and steps.

Step 1: Assess Power Requirements

1. Calculate total wattage:
   - Add up the wattage of all infrared panels
   - Include any additional features (lights, audio systems, etc.)

2. Determine voltage requirements:
   - Most home infrared saunas use 120V or 240V
   - Higher voltage typically allows for more powerful systems

3. Calculate amperage:
   - Use the formula: Amps = Watts ÷ Volts
   - Ensure your electrical system can handle the load

Step 2: Plan the Electrical Layout

1. Sketch a wiring diagram:
   - Show panel locations
   - Include control panel and power source
   - Plan wire routes

2. Choose appropriate wire gauge:
   - Based on amperage and distance
   - Consult electrical code or an electrician

3. Plan for a dedicated circuit:
- Typically 20-30 amp circuit for most home saunas
- May require 40-50 amp for larger or commercial units

Step 3: Gather Materials and Tools

- Appropriate gauge electrical wire
- Circuit breaker
- Junction boxes
- Conduit (if required by local code)
- Wire nuts and electrical tape
- Voltage tester
- Screwdrivers and wire strippers
- Safety equipment (gloves, safety glasses)

Step 4: Install the Dedicated Circuit

CAUTION: This step should be performed by a licensed electrician.

1. Install new circuit breaker in main panel
2. Run appropriate wire from panel to sauna location
3. Install junction box near sauna for power connection

Step 5: Prepare for Panel Wiring

1. Turn off power at the main breaker
2. Install electrical boxes for each panel location
3. Run wires from the main junction box to each panel location:
- Use conduit if required by local code
- Ensure proper wire length at each location

Step 6: Wire the Control Panel

1. Mount the control panel in desired location:
- Usually outside the sauna for temperature control
2. Connect control panel to power supply
3. Run low-voltage control wires to each infrared panel

Step 7: Connect Infrared Panels

1. Follow manufacturer's wiring diagram for each panel
2. Make connections in junction boxes:
- Strip wire ends carefully
- Use wire nuts for secure connections
- Wrap connections with electrical tape for added security

3. Secure all wiring and close junction boxes

Step 8: Install Safety Features

1. Add a GFCI (Ground Fault Circuit Interrupter) if required:
- Protects against ground faults
- May be built into the circuit breaker or installed separately

2. Ensure proper grounding of all components

Step 9: Final Connections and Testing

1. Double-check all connections
2. Close all electrical boxes and panels
3. Turn power on at the main breaker
4. Test each component of the sauna system

Troubleshooting Tips:

1. Circuit Breaker Trips:
- Verify total amperage doesn't exceed circuit capacity
- Check for short circuits or loose connections
- Consider upgrading to a higher amperage circuit if necessary

2. Panels Not Heating:
- Check connections at each panel
- Verify control panel is sending proper signals
- Test voltage at panel connections

3. Inconsistent Heating:
- Ensure all panels are on the same circuit
- Check for loose connections
- Verify control panel is functioning correctly

4. Control Panel Not Responding:
- Check power supply to control panel
- Verify low-voltage connections to panels
- Consider resetting or replacing the control panel

5. Flickering Lights:
- Ensure lights are on a separate circuit from heating panels
- Check for loose connections
- Verify proper voltage for light fixtures

6. Electrical Humming or Buzzing:
- Tighten all connections
- Check for any wires touching metal surfaces
- Verify proper grounding of all components

7. Partial Power to Sauna:
- Check for tripped GFCI
- Verify all wires are properly connected in junction boxes
- Test voltage at various points to isolate the issue

8. Overheating:
- Verify temperature sensor is properly connected and placed
- Check control panel settings
- Ensure ventilation is adequate

9. GFCI Constantly Tripping:
- Check for moisture in electrical boxes
- Verify insulation on all wires is intact
- Consider replacing the GFCI if issue persists

10. WiFi Control Issues:
- Ensure strong WiFi signal in sauna area
- Update control panel firmware if available
- Verify compatibility between app and control system

Remember, electrical work can be dangerous and may require specific certifications depending on your local regulations. While understanding these principles is valuable, it's often best to hire a licensed electrician for the actual installation, especially for tasks involving the main electrical panel or complex wiring.

Are you feeling more confident about the electrical considerations for your infrared sauna? Which aspects of the electrical setup do you think will be most crucial for your specific sauna design? Remember, prioritizing safety and following local electrical codes is paramount when dealing with any electrical installation.

Customizing for Optimal Performance

Customizing your infrared sauna for optimal performance ensures you get the most out of your investment and enjoy a tailored, efficient sauna experience. Let's explore the steps to fine-tune your infrared sauna for peak performance.

Step 1: Analyze Your Sauna Usage

1. Identify primary users and their preferences:
   - Temperature ranges
   - Session durations
   - Favorite times of use

2. List desired health benefits:
   - Relaxation
   - Pain relief
   - Detoxification
   - Cardiovascular health

Step 2: Optimize Panel Placement

1. Reassess panel locations:
   - Ensure even coverage for seated users
   - Consider adding panels for problem areas (e.g., feet, shoulders)

2. Adjust panel angles:
   - Aim panels towards the body for maximum effectiveness
   - Use adjustable brackets for flexibility

Step 3: Enhance Temperature Control

1. Install a smart thermostat:
 - Allows for precise temperature control
 - Enables pre-heating and scheduling

2. Add zoned heating:
 - Install separate controls for different areas (e.g., back, legs)
 - Allows customization for multiple users

Step 4: Improve Comfort Features

1. Upgrade seating:
 - Install ergonomic benches
 - Add removable backrests for comfort

2. Enhance lighting:
 - Install chromotherapy lights for added benefits
 - Use dimmable LEDs for ambiance control

Step 5: Optimize Air Quality

1. Install a ventilation system:
 - Small, quiet fan for air circulation
 - Helps maintain comfortable humidity levels

2. Add an air purification system:
 - HEPA filter to remove airborne particles
 - Enhances overall sauna experience

Step 6: Incorporate Audio/Visual Elements

1. Install a sound system:
   - Bluetooth speakers for music or guided meditations
   - Ensure components are heat-resistant

2. Add a tablet mount or TV:
   - For entertainment or educational content during sessions
   - Use heat-resistant mounting solutions

Step 7: Enhance Relaxation Features

1. Install aromatherapy diffusers:
   - Use heat-safe essential oil dispensers
   - Integrate with ventilation system if possible

2. Add salt walls or bricks:
   - Provides additional health benefits
   - Enhances aesthetic appeal

Step 8: Implement Smart Features

1. Install WiFi-enabled controls:
   - Allow remote operation and monitoring
   - Enable integration with smart home systems

2. Add usage tracking:
   - Monitor session durations and frequencies
   - Track energy consumption for efficiency

Step 9: Personalize the Aesthetics

1. Customize interior finishes:
   - Choose wood types or colors that suit your style
   - Add decorative elements that can withstand heat

2. Enhance exterior design:
- Match sauna exterior to your home decor
- Consider custom door designs or window placements

Step 10: Fine-tune and Test

1. Conduct multiple test sessions:
- Adjust settings for different scenarios
- Gather feedback from various users

2. Create preset programs:
- Design programs for specific goals (e.g., relaxation, detox)
- Save favorite settings for quick access

Troubleshooting Tips:

1. Uneven Heating:
- Reposition panels or add additional panels to cold spots
- Check for any obstructions blocking infrared rays
- Verify all panels are functioning at full capacity

2. Difficulty Reaching Desired Temperature:
- Improve insulation, especially around the door and windows
- Verify power supply is adequate for all panels
- Consider pre-heating the sauna before use

3. Excessive Heat in Specific Areas:
- Adjust panel angles or distance from seating
- Implement zoned heating controls
- Add heat-reflective material behind problem areas

4. Poor Air Quality:
- Increase ventilation rate
- Clean or replace air filters regularly
- Ensure proper sealing to prevent outside air infiltration

5. Inconsistent Performance:
- Check electrical connections for loose wires
- Update control system firmware if available
- Consider a power conditioner for stable electrical supply

6. Discomfort During Sessions:
- Adjust seating position or upgrade ergonomic features
- Implement gradual heating programs for acclimation
- Ensure proper hydration before and after sessions

7. Condensation Issues:
- Improve ventilation to reduce humidity
- Check for any cold spots in sauna construction
- Consider adding a small dehumidifier for extreme cases

8. Inefficient Energy Use:
- Implement scheduling to avoid unnecessary heating
- Use reflective insulation to maximize infrared efficiency
- Ensure door and windows are properly sealed

9. Technology Integration Problems:
- Ensure strong WiFi signal in sauna area
- Keep software and apps updated
- Consider a dedicated router for smart sauna features

10. Overwhelming Features:
- Create simple, one-touch programs for easy use
- Provide clear instructions for all features
- Consider removing rarely used features for simplicity

Remember, customizing your infrared sauna is an ongoing process. As you use your sauna more, you'll discover what works best for you and can continue to refine the experience. Don't be afraid to experiment with different settings and features to find your perfect sauna setup.

Are you excited about personalizing your infrared sauna experience? Which customization features are you most looking forward to implementing? Remember, the goal is to create a sauna that not only meets your health and wellness needs but also provides a truly enjoyable and relaxing experience tailored to your preferences.

Chapter 7
Sauna Accessories and Enhancements

Seating Options and Ergonomics

Comfortable and ergonomic seating is crucial for an enjoyable sauna experience. Let's explore how to create the perfect seating arrangement for your sauna, focusing on comfort, functionality, and durability.

Step 1: Assess Your Space and Needs

1. Measure your sauna interior:
   - Available floor space
   - Ceiling height
   - Door and heater locations

2. Determine user requirements:
   - Number of regular users
   - Any specific health needs (e.g., back support, accessibility)
   - Preferred sauna positions (sitting, reclining, lying down)

Step 2: Choose Bench Style

1. Traditional straight benches:
   - Simple and space-efficient
   - Good for smaller saunas

2. L-shaped benches:
   - Maximizes corner space
   - Allows for lying down in longer saunas

3. U-shaped benches:
   - Ideal for larger saunas
   - Great for social sauna experiences

4. Tiered benches:
   - Allows for temperature variation
   - Maximizes vertical space

Step 3: Select Bench Material

1. Western Red Cedar:
   - Classic choice, naturally resistant to decay
   - Pleasant aroma

2. Hemlock:
   - Light color, minimal scent
   - Durable and splinter-resistant

3. Abachi:
   - Very low heat conductivity
   - Comfortable for bare skin

4. Basswood:
   - Light color, virtually knot-free
   - Soft and comfortable

Step 4: Determine Bench Dimensions

1. Depth:
   - 18-24 inches for comfortable sitting
   - 24-28 inches if planning for reclining

2. Width:
   - Allow 24-28 inches per person

3. Height:
- Lower bench: 18-20 inches from the floor
- Upper bench: 36-40 inches from the floor

4. Spacing:
- 12-14 inches between upper and lower bench for leg room

Step 5: Incorporate Ergonomic Features

1. Curved seat edges:
- Reduces pressure on the back of thighs

2. Sloped backrest:
- 100-110 degree angle for optimal comfort

3. Headrests:
- Removable for easy cleaning
- Adjustable for different heights

4. Footrests:
- Helps shorter users reach optimal heat levels

Step 6: Add Comfort Enhancements

1. Bench skirts:
- Prevents heat from rising between bench slats

2. Removable bench covers:
- Provides a cooler seating surface
- Easy to wash and maintain

3. Lumbar supports:
- Adjustable for personalized comfort

Step 7: Consider Accessibility

1. Install grab bars:
   - Aids in moving between bench levels
   - Helpful for users with mobility issues

2. Add a flip-up bench:
   - Creates space for wheelchair users
   - Useful for changing or cool-down area

Step 8: Build and Install

1. Cut lumber to size:
   - Use stainless steel screws to prevent rusting

2. Assemble bench frames:
   - Ensure sturdy construction

3. Attach bench tops:
   - Leave small gaps between slats for air circulation

4. Secure benches to wall studs:
   - Use brackets for added stability

5. Sand all surfaces thoroughly:
   - Round edges for comfort and safety

Step 9: Test and Adjust

1. Sit on all benches to test comfort
2. Lie down (if space allows) to check for adequate length
3. Ensure stability of all seating elements

Troubleshooting Tips:

1. Uncomfortable Seating:
- Add cushions or bench covers for softer surface
- Adjust backrest angle for better support
- Consider replacing bench tops with a different wood type

2. Unstable Benches:
- Reinforce bench frames with additional supports
- Check and tighten all screws and brackets
- Ensure proper attachment to wall studs

3. Overheating on Upper Benches:
- Install heat deflectors above seating area
- Add removable insulating covers for bench tops
- Adjust heater output or placement if possible

4. Difficulty Moving Between Bench Levels:
- Install additional grab bars or handholds
- Add intermediate step between upper and lower benches
- Consider a small, removable ladder for easier access

5. Benches Too Hard:
- Introduce thin, heat-resistant cushions
- Try different wood types known for softness (e.g., Abachi)
- Sand bench surfaces to ultra-smooth finish

6. Condensation on Bench Surfaces:
- Improve ventilation in the sauna
- Use woods with better moisture-resistance properties
- Apply a thin coat of water-based sealant (ensure it's sauna-safe)

7. Benches Too Short for Lying Down:
- Install fold-down extensions at bench ends
- Consider removable leg supports for extended seating
- Redesign bench layout to accommodate longer sections

8. Splinters or Rough Spots:
- Sand problem areas thoroughly
- Apply food-grade mineral oil to wood surface
- Replace damaged wood sections if necessary

9. Inadequate Lower Back Support:
- Install adjustable lumbar cushions
- Modify backrest angle for better ergonomics
- Add removable rolled towels for customizable support

10. Benches Too High/Low for Comfort:
- Adjust bench heights during installation
- For existing benches, consider adding removable platforms or footrests
- Install adjustable-height benches for maximum flexibility

Remember, comfortable seating is key to a enjoyable and relaxing sauna experience. Take the time to design and implement seating that meets your specific needs and preferences. Don't hesitate to make adjustments based on user feedback to create the perfect ergonomic setup for your sauna.

Are you excited about designing the ideal seating arrangement for your sauna? Which ergonomic features do you think will be most important for your comfort? Remember, investing time and thought into your sauna's seating will greatly enhance your overall sauna experience and encourage more frequent use.

Thermometers and Hygrometers

Accurate temperature and humidity measurement are crucial for maintaining a safe and effective sauna environment. Let's explore how to select, install, and use thermometers and hygrometers in your sauna.

Step 1: Understanding the Importance

1. Temperature monitoring:
   - Ensures safe operating conditions
   - Helps achieve desired therapeutic effects

2. Humidity tracking (for traditional saunas):
   - Affects perceived heat and comfort
   - Influences the sauna's health benefits

Step 2: Choosing the Right Devices

1. Thermometer options:
   - Analog: Classic look, no batteries needed
   - Digital: Precise readings, often with additional features
   - Infrared: Non-contact temperature measurement

2. Hygrometer options:
   - Analog: Simple and reliable
   - Digital: More accurate, often combined with thermometers

3. Combo devices:
   - Thermo-hygrometers: Measure both temperature and humidity
   - Often come with additional features like min/max memory

Step 3: Selecting Sauna-Specific Devices

Look for:
1. Heat resistance (up to at least 200°F/93°C)
2. Moisture resistance
3. Accuracy in high-temperature environments
4. Easy-to-read display (large numbers, backlight for digital devices)

Step 4: Determining Placement

1. Temperature measurement:
- Place thermometer at head level when seated
- Typically 5-6 feet from the floor
- Away from direct heat sources and cold air inlets

2. Humidity measurement:
- Place hygrometer at similar height as thermometer
- Away from direct steam or water sources

3. Consider multiple measurement points for larger saunas

Step 5: Installation Process

1. For wooden walls:
- Use small stainless steel screws or nails
- Pre-drill holes to prevent wood splitting

2. For tile or stone surfaces:
- Use heat-resistant adhesive or mounting tape
- Ensure the surface is clean and dry before mounting

3. For digital devices:
- Follow manufacturer's instructions for probe placement
- Ensure any wiring is heat-resistant and properly secured

Step 6: Calibration and Testing

1. Allow devices to acclimate to the sauna environment
2. Compare readings with a known accurate device
3. Calibrate if necessary (refer to device manual)
4. Test at various temperatures and humidity levels

Step 7: Regular Maintenance

1. Clean devices regularly with a damp cloth
2. Check battery life for digital devices
3. Verify accuracy periodically
4. Replace devices if they show signs of deterioration or inaccuracy

Step 8: Using Your Devices Effectively

1. Check temperature before entering the sauna
2. Monitor changes throughout your session
3. Use readings to adjust sauna settings for optimal experience
4. Keep a log to track your preferred conditions

Troubleshooting Tips:

1. Inconsistent Readings:
   - Verify proper placement away from heat sources
   - Check for drafts or cold spots affecting the device
   - Compare with a secondary device to confirm accuracy

2. Foggy or Condensation-Covered Display:
   - Improve ventilation in the sauna
   - Consider a device with better moisture resistance
   - Use an anti-fog treatment on analog displays

3. Digital Device Malfunction in High Heat:
 - Ensure the device is rated for sauna temperatures
 - Place the display unit outside the sauna with only the probe inside
 - Consider switching to an analog device

4. Difficulty Reading Analog Devices:
 - Install better lighting near the device
 - Choose a device with larger, high-contrast markings
 - Consider replacing with a digital device with a backlit display

5. Hygrometer Showing 100% Constantly:
 - Check for direct exposure to steam or water
 - Verify the device hasn't been damaged by excessive moisture
 - Calibrate or replace if necessary

6. Delayed Response to Temperature Changes:
 - Ensure the device isn't insulated from the air
 - Look for devices with faster response times
 - For digital devices, check that the probe is properly exposed

7. Battery Life Issues in Digital Devices:
 - Use high-quality, heat-resistant batteries
 - Consider devices with external power options
 - Switch to analog devices for maintenance-free operation

8. Discrepancies Between Multiple Devices:
 - Verify all devices are properly calibrated
 - Account for normal temperature stratification in the sauna
 - Use the average of multiple readings for best accuracy

9. Difficulty Mounting on Certain Surfaces:
- Use heat-resistant mounting brackets or stands
- Consider freestanding devices for flexible placement
- For glass walls, use specialized suction cup mounts

10. Inaccurate Humidity Readings in Dry Saunas:
- Ensure the hygrometer is designed for low-humidity environments
- Calibrate using a salt test kit
- Consider a more sensitive digital hygrometer

Remember, accurate temperature and humidity monitoring is essential for a safe and effective sauna experience. Regular checks and maintenance of your thermometer and hygrometer will help ensure you're always enjoying your sauna at optimal conditions.

Are you feeling confident about selecting and installing the right temperature and humidity monitoring devices for your sauna? Which features do you think will be most important for your sauna use? Remember, while these devices are crucial for safety, they also help you fine-tune your sauna experience to your personal preferences, enhancing your overall enjoyment and benefit from sauna sessions.

Aromatherapy and Essential Oils

Incorporating aromatherapy into your sauna experience can enhance relaxation, promote wellness, and create a more immersive sensory experience. Let's explore how to safely and effectively use essential oils in your sauna.

Step 1: Understanding Aromatherapy in Saunas

1. Benefits:
   - Enhances relaxation
   - Promotes respiratory health
   - Can boost mood and reduce stress
   - Adds a personalized touch to your sauna

2. Safety considerations:
   - Heat intensifies oil potency
   - Some oils may irritate skin or airways
   - Quality and purity of oils are crucial

Step 2: Choosing Essential Oils

1. Relaxation: Lavender, Chamomile, Ylang-Ylang
2. Respiratory support: Eucalyptus, Peppermint, Tea Tree
3. Mood-boosting: Citrus oils (Orange, Lemon), Bergamot
4. Muscle relief: Rosemary, Marjoram, Ginger
5. Skin health: Geranium, Frankincense, Sandalwood

Step 3: Selecting a Diffusion Method

1. Water bucket method:
   - Add a few drops of oil to water for ladling over hot stones

2. Essential oil diffuser:
- Choose a heat-resistant, battery-operated model

3. Aromatic sprays:
- Create a pre-mixed spray to mist in the sauna

4. Infused wooden accessories:
- Apply oils to cedar blocks or other wooden items

Step 4: Preparing Your Aromatherapy Setup

1. For water bucket method:
- Use a ceramic or wooden bucket
- Prepare a mix of 2-3 drops of oil per cup of water

2. For diffusers:
- Place in a safe location away from direct heat
- Follow manufacturer's instructions for oil-to-water ratio

3. For aromatic sprays:
- Mix 10-15 drops of oil with 1 oz of witch hazel in a spray bottle
- Top up with distilled water

4. For wooden accessories:
- Apply 5-10 drops of oil to a cedar block or wooden bowl

Step 5: Safe Application in the Sauna

1. Start with less:
- Begin with 1-2 drops and increase gradually

2. Timing:
- Add oils at the beginning of your session for a lasting effect

3. Ventilation:
 - Ensure proper air circulation to prevent overwhelming scents

4. Personal sensitivity:
 - Be aware of any allergies or sensitivities

Step 6: Creating Blends

1. Choose complementary scents:
 - Example: Lavender + Eucalyptus for relaxation and clear breathing

2. Balance ratios:
 - Use 2 parts base note, 1 part middle note, 1 part top note

3. Test blends before use:
 - Mix small amounts to find your preferred combination

Step 7: Maintaining Your Aromatherapy Setup

1. Clean diffusers regularly:
 - Wipe with alcohol between uses

2. Store oils properly:
 - Keep in dark, cool place away from sauna heat

3. Replace oils as needed:
 - Most oils last 1-2 years when stored properly

Step 8: Experimenting and Personalizing

1. Keep a journal:
 - Note which scents you enjoy and their effects

2. Rotate oils:
- Change scents based on mood or desired benefits

3. Seasonal adjustments:
- Use invigorating scents in morning, calming at night

Troubleshooting Tips:

1. Overwhelming Scent:
- Reduce the amount of oil used
- Improve ventilation in the sauna
- Try a milder oil or blend

2. Skin Irritation:
- Discontinue use of the current oil
- Ensure oils are properly diluted
- Try patch testing oils before use in the sauna

3. Respiratory Discomfort:
- Switch to gentler oils like Lavender or Chamomile
- Reduce the amount of oil used
- Ensure adequate ventilation

4. Oil Not Diffusing Properly:
- Check that your diffuser is heat-safe and functioning
- Clean the diffuser thoroughly
- Try a different diffusion method

5. Scent Fading Too Quickly:
- Use base notes like Sandalwood or Cedarwood for longer-lasting scents
- Reapply oils halfway through your session
- Try a higher-quality, more concentrated oil

6. Oils Damaging Sauna Wood:
- Always dilute oils before applying directly to wood
- Use a separate ceramic or glass dish for oil application
- Consider using pre-infused wooden accessories

7. Difficulty Blending Scents:
- Start with simple two-oil blends
- Follow the 30-50-20 rule: 30% top note, 50% middle note, 20% base note
- Keep notes on successful and unsuccessful combinations

8. Oils Evaporating Too Fast:
- Store oils outside the sauna in a cool, dark place
- Use smaller amounts more frequently
- Try oils with lower volatility (like base notes)

9. Allergic Reactions:
- Immediately discontinue use and ventilate the sauna
- Always do a patch test before introducing new oils
- Consider consulting with an aromatherapist for safe alternatives

10. Conflicting Scents with Multiple Users:
- Establish a neutral scent everyone enjoys
- Use personal inhalers for individual aromatherapy experiences
- Create a schedule for different scents on different days

Remember, aromatherapy should enhance your sauna experience, not overpower it. Start conservatively and adjust based on your preferences and reactions. Always prioritize safety and be mindful of others who may use the sauna.

Are you excited to incorporate aromatherapy into your sauna routine? Which essential oils or blends are you most looking forward to trying? Remember, the right scent can transform your sauna session, making it a more relaxing, invigorating, or therapeutic experience tailored to your needs and preferences.

Audio Systems and Entertainment

Integrating audio systems and entertainment options into your sauna can greatly enhance your relaxation experience. Let's explore how to safely and effectively add these features to your sauna environment.

Step 1: Assess Your Entertainment Needs

1. Determine your preferences:
   - Music only
   - Podcasts or audiobooks
   - Video capabilities
   - Meditation or guided relaxation audio

2. Consider sauna usage:
   - Solo or multiple users
   - Frequency and duration of sessions

Step 2: Choose Your Audio System

1. Bluetooth speakers:
   - Wireless, easy to install
   - Can be controlled from outside the sauna

2. Built-in stereo system:
   - Integrated look
   - Better sound quality

3. Waterproof portable speakers:
   - Flexible placement
   - Can be removed when not in use

Step 3: Select Heat-Resistant Components

1. Look for:
- Operating temperature range up to 175°F (80°C)
- Moisture-resistant or waterproof ratings (IPX5 or higher)
- Corrosion-resistant materials

2. Consider:
- Marine-grade speakers
- Sauna-specific audio systems

Step 4: Plan Your Layout

1. Speaker placement:
- Above or below benches
- In corners for better sound distribution
- Away from direct heat sources

2. Control panel location:
- Outside the sauna for easy access
- Near the door for convenience

3. Wiring routes:
- Behind walls or under benches
- Use heat-resistant wiring

Step 5: Installation Process

1. For built-in systems:
- Cut openings for speakers and control panel
- Install vapor barrier around all cuts
- Mount speakers and seal edges

2. For portable or Bluetooth systems:
- Create a stable, heat-safe mounting area
- Ensure easy access for controls and charging

3. Wiring (if applicable):
- Use heat-resistant, insulated wiring
- Secure wires away from heat sources
- Seal all penetrations in sauna walls

Step 6: Set Up Entertainment Sources

1. Bluetooth pairing:
- Sync with your smartphone or tablet

2. Media server:
- Set up a dedicated device for audio storage

3. Streaming services:
- Ensure good WiFi coverage in sauna area

Step 7: Enhance the Experience

1. Create sauna playlists:
- Relaxing music
- Nature sounds
- Guided meditations

2. Install mood lighting:
- Sync with audio for immersive experience

3. Consider aromatherapy integration:
- Pair scents with specific playlists

Step 8: Maintain Your System

1. Regular cleaning:
- Wipe down components with a slightly damp cloth

2. Check connections:
- Ensure all wires and mounts are secure

3. Update software:
- Keep firmware current for best performance

Troubleshooting Tips:

1. Poor Sound Quality:
- Adjust speaker placement for better acoustics
- Check for loose connections
- Consider adding a small subwoofer for better bass response

2. Bluetooth Connectivity Issues:
- Ensure device is within range
- Reset Bluetooth connection
- Update device and speaker firmware

3. Moisture Damage:
- Improve sauna ventilation
- Use a dehumidifier in the sauna when not in use
- Consider more water-resistant components

4. Heat-Related Malfunctions:
- Verify all components are rated for sauna temperatures
- Install heat shields around sensitive equipment
- Consider moving amplifiers or control units outside the sauna

5. Interference with Relaxation:
- Lower volume or use directional speakers
- Create separate zones for those who prefer quiet
- Use wireless headphones for personal listening

6. WiFi Signal Weakness:
- Install a WiFi extender near the sauna
- Consider a mesh network system for better coverage
- Use offline playlists when streaming is unreliable

7. Battery Life Issues (for portable devices):
- Keep devices charged between sessions
- Install a heat-safe charging station in the sauna
- Use devices with longer battery life or wired options

8. Corrosion on Connections:
- Use corrosion-resistant connectors and terminals
- Apply dielectric grease to connections
- Regularly inspect and clean connections

9. Volume Fluctuations:
- Check for loose wires or connections
- Ensure amplifier is properly rated for the environment
- Consider a volume stabilizer or limiter

10. System Overheating:
- Improve ventilation around audio components
- Use heat sinks on amplifiers if necessary
- Consider a cooling system for high-powered setups

Remember, while entertainment can enhance your sauna experience, it's important to balance it with the traditional benefits of quiet relaxation. Always prioritize safety when installing electrical components in your sauna, and be mindful of creating an environment that promotes overall wellness.

Are you excited about adding audio or entertainment features to your sauna? What type of content do you think will best enhance your sauna sessions? Remember, the right audio setup can transform your sauna into a multi-sensory relaxation chamber, helping you unwind and de-stress more effectively. Whether you prefer soothing music, informative podcasts, or guided meditations, tailoring your audio experience can significantly boost the benefits of your sauna time.

Chapter 8
Maintenance and Care

Cleaning Procedures for Different Sauna Types

Proper cleaning and maintenance are crucial for ensuring the longevity, hygiene, and optimal performance of your sauna. Let's explore the cleaning procedures for different sauna types: traditional dry saunas, steam saunas, and infrared saunas.

General Cleaning Guidelines:

1. Frequency:
   - Light cleaning after each use
   - Deep cleaning weekly or bi-weekly, depending on usage

2. Basic supplies:
   - Soft brush or cloth
   - Mild, natural soap (avoid harsh chemicals)
   - White vinegar
   - Baking soda
   - Bucket of warm water
   - Microfiber cloths

Step-by-Step Cleaning Process:

A. Traditional Dry Sauna

1. Daily/After Each Use:
   - Remove any visible debris
   - Wipe down benches with a damp cloth
   - Leave door open to air out

2. Weekly Deep Clean:
a. Prepare cleaning solution:
- Mix 1 cup white vinegar with 1 gallon warm water

b. Clean wooden surfaces:
- Dip soft brush in solution and gently scrub benches and walls
- Pay extra attention to areas with body contact

c. Rinse and dry:
- Wipe down with clean, damp cloth
- Dry thoroughly with microfiber cloth

d. Clean floor:
- Sweep or vacuum, then mop with vinegar solution

e. Clean door and windows:
- Use vinegar solution on glass surfaces

f. Disinfect high-touch areas:
- Door handles, thermostats, etc.

3. Monthly Maintenance:
- Sand any rough spots on wooden surfaces
- Check and clean heater rocks

B. Steam Sauna

1. Daily/After Each Use:
- Wipe down all surfaces with a clean, damp cloth
- Run ventilation fan to reduce moisture

2. Weekly Deep Clean:
a. Prepare cleaning solution:
- Mix equal parts water and white vinegar

b. Clean tiles and grout:
- Spray solution on surfaces
- Scrub with a soft brush
- Rinse thoroughly with clean water

c. Clean and disinfect benches:
- Use vinegar solution or a sauna-safe disinfectant

d. Descale steam generator:
- Follow manufacturer's instructions for descaling

e. Clean drains:
- Use a mixture of baking soda and vinegar to clear drains

3. Monthly Maintenance:
- Check and clean ventilation systems
- Inspect grout and reseal if necessary

C. Infrared Sauna

1. Daily/After Each Use:
- Wipe down benches and backrest with a damp cloth
- Clean any glass surfaces

2. Weekly Deep Clean:
a. Prepare cleaning solution:
- Mix mild soap with warm water

b. Clean wooden surfaces:
- Wipe down all surfaces with soapy solution
- Rinse with a clean, damp cloth
- Dry thoroughly

c. Clean infrared emitters:
- Gently dust or wipe with a soft, dry cloth

d. Clean controls and display:
- Use a slightly damp microfiber cloth

e. Vacuum floor and corners

3. Monthly Maintenance:
- Check all electrical connections
- Inspect emitters for any damage

Troubleshooting Tips:

1. Persistent Odors:
- Place a bowl of white vinegar in the sauna overnight
- Use a natural odor absorber like activated charcoal
- Check for hidden mold or mildew

2. Mold or Mildew Growth:
- Improve ventilation
- Use a natural anti-fungal solution (tea tree oil diluted in water)
- Sand and refinish affected wood if necessary

3. Stained Wood:
- Try a mixture of baking soda and water for light stains
- For tougher stains, lightly sand the area
- Avoid bleach or harsh chemicals

4. Cloudy Glass:
- Use a mixture of equal parts vinegar and water
- For mineral buildup, try a commercial lime scale remover

5. Sticky Residue:
- Use a mixture of baking soda and coconut oil as a gentle abrasive
- For stubborn residue, try a small amount of mineral oil

6. Scratched Surfaces:
- For light scratches, try rubbing with a walnut (the oils can help)
- Sand lightly and apply a food-grade mineral oil

7. Malfunctioning Controls (Infrared):
- Ensure cleaning solution doesn't seep into control panels
- Use only a slightly damp cloth for cleaning electronics

8. Scale Buildup in Steam Generator:
- Use a vinegar solution for mild buildup
- For heavy buildup, use a commercial descaling product designed for saunas

9. Drain Clogs:
- Use a plumber's snake for stubborn clogs
- Regularly use enzymatic drain cleaners to prevent buildup

10. Faded or Discolored Wood:
- Apply a UV-resistant wood oil annually
- Consider using a sauna-safe wood stain to refresh the color

Remember, regular maintenance and prompt attention to any issues will keep your sauna in top condition for years to come. Always use products specifically designed for saunas or natural alternatives to avoid introducing harmful chemicals into your sauna environment.

Are you feeling confident about maintaining your specific type of sauna? Which cleaning task do you think will be most important for preserving your sauna's quality and hygiene? Remember, a well-maintained sauna not only looks and smells better but also provides a healthier, more enjoyable experience for you and your guests.

Wood Treatment and Preservation

Proper wood treatment and preservation are essential for maintaining the beauty, functionality, and longevity of your sauna. Let's explore the steps to keep your sauna's wood in top condition.

Step 1: Understanding Your Sauna Wood

1. Identify the wood type:
    - Common types: Western Red Cedar, Hemlock, Nordic White Spruce
    - Each wood type has unique characteristics and care needs

2. Assess the current condition:
    - Look for signs of wear, discoloration, or damage
    - Note any areas of concern (e.g., benches, high-traffic zones)

Step 2: Regular Cleaning

1. Daily/After each use:
    - Wipe down surfaces with a damp cloth
    - Use only water, avoid soaps or chemicals

2. Weekly:
    - Soft brush cleaning with mild, natural soap solution
    - Rinse thoroughly and dry completely

Step 3: Sanding

1. When to sand:
    - If wood feels rough or splinters are present
    - Before applying new treatments

2. Sanding process:
- Use fine-grit sandpaper (180-220 grit)
- Sand in the direction of the wood grain
- Remove all dust with a vacuum and tack cloth

Step 4: Choosing Wood Treatment Products

1. For untreated wood:
- Heat-resistant, food-grade mineral oil
- Paraffin oil
- Natural beeswax

2. For pre-treated wood:
- Sauna-specific wood treatment products
- Always check manufacturer recommendations

Step 5: Applying Wood Treatment

1. Preparation:
- Ensure wood is clean and completely dry
- Test product on a small, inconspicuous area first

2. Application process:
- Apply a thin, even coat using a lint-free cloth
- Work in the direction of the wood grain
- Pay extra attention to end grains and high-wear areas

3. Drying and curing:
- Allow to dry completely (usually 24-48 hours)
- Ensure good ventilation during drying process

Step 6: Sealing and Protecting

1. Consider applying a heat-resistant sealant:
   - Helps protect against moisture and stains
   - Use only products specifically designed for saunas

2. Apply sealant:
   - Use a natural bristle brush or lint-free cloth
   - Apply thin, even coats
   - Allow each coat to dry completely before reapplication

Step 7: Ongoing Maintenance

1. Reapplication schedule:
   - Benches and high-traffic areas: Every 6-12 months
   - Walls and ceiling: Every 1-2 years

2. Regular inspections:
   - Check for signs of wear or damage monthly
   - Address issues promptly to prevent further damage

Step 8: UV Protection (for saunas with windows)

1. Apply UV-resistant treatment to exposed areas
2. Consider installing UV-filtering film on windows

Troubleshooting Tips:

1. Persistent Dryness:
   - Increase treatment frequency
   - Use a higher-quality, more penetrating oil
   - Check sauna humidity levels and adjust if necessary

2. Sticky Surfaces After Treatment:
- You may have applied too much product
- Gently wipe excess with a clean, dry cloth
- Allow more drying time between coats

3. Uneven Color or Finish:
- Sand lightly and reapply treatment
- Ensure even application with long, smooth strokes
- Consider using a spray application for more uniformity

4. Mold or Mildew Growth:
- Improve ventilation in the sauna
- Clean affected areas with a vinegar solution
- Apply a natural anti-fungal treatment before re-oiling

5. Wood Turning Gray or Black:
- This is often due to UV exposure or mineral deposits
- Sand affected areas and reapply UV-resistant treatment
- Consider using a wood brightener before retreatment

6. Cracking or Splitting Wood:
- This can be due to excessive dryness
- Apply a deep-penetrating oil treatment
- Maintain consistent humidity levels in the sauna

7. Difficulty in Maintaining Benches:
- Use removable bench covers for easier cleaning and maintenance
- Apply an extra coat of treatment to high-wear areas
- Consider using a harder wood type for bench replacement

8. Strong Odor After Treatment:
- Ensure you're using sauna-safe, low-VOC products
- Increase ventilation after application
- Allow longer curing time before using the sauna

9. Water Stains:
- Sand lightly to remove the stain
- Apply a wood bleach if necessary, then retreat
- Improve water resistance with an appropriate sealant

10. Resin Seepage (for cedar saunas):
- This is natural but can be messy
- Gently scrape away excess resin
- Apply heat (carefully) to crystallize and remove remaining resin
- Treat the area with a stain-blocking primer before re-oiling

Remember, the key to successful wood treatment and preservation is consistency and using the right products. Always prioritize sauna-safe, natural treatments over harsh chemicals. Proper care will not only keep your sauna looking beautiful but also ensure a safe and hygienic environment for relaxation.

Are you feeling prepared to tackle the wood treatment in your sauna? Which aspect of wood care do you think will be most crucial for maintaining your sauna's quality? Remember, a well-maintained wooden sauna not only looks and feels better but can also provide a more authentic and enjoyable sauna experience for years to come.

Troubleshooting Common Issues

Even with regular maintenance, sauna owners may encounter various issues. Let's explore common problems and their solutions to keep your sauna in top condition.

1. Sauna Not Heating Properly

Symptoms:
- Slow to reach desired temperature
- Not reaching set temperature
- Uneven heating

Troubleshooting steps:
a) Check power supply:
   - Ensure circuit breaker hasn't tripped
   - Verify all connections are secure

b) Inspect heating elements:
   - Look for visible damage or corrosion
   - Test continuity with a multimeter

c) Examine thermostat:
   - Calibrate if readings are off
   - Replace if malfunctioning

d) Check for air leaks:
   - Inspect door seals
   - Look for gaps in walls or ceiling

Solution:
- Replace faulty components
- Seal any air leaks
- Call a professional for electrical issues

2. Excessive Humidity (Traditional Saunas)

Symptoms:
- Difficulty breathing
- Water condensation on walls
- Mold growth

Troubleshooting steps:
a) Check ventilation:
  - Ensure vents are open and unobstructed
  - Clean any dirty vents or fans

b) Inspect for water leaks:
  - Look for wet spots on walls or floor
  - Check integrity of vapor barrier

c) Assess sauna usage:
  - Avoid excessive water on stones
  - Allow proper drying time between sessions

Solution:
- Improve ventilation (add a fan if necessary)
- Fix any leaks and reinforce waterproofing
- Use a dehumidifier between sessions if needed

3. Wood Damage

Symptoms:
- Cracks in benches or walls
- Warping of wood panels
- Discoloration

Troubleshooting steps:
a) Identify cause:
- Check for water damage
- Assess heat exposure
- Look for signs of pests

b) Evaluate wood condition:
- Determine if wood can be repaired or needs replacement
- Check structural integrity

c) Review maintenance practices:
- Ensure proper cleaning and oiling routines
- Verify correct humidity levels

Solution:
- Sand and refinish minor damage
- Replace severely damaged wood
- Adjust maintenance routine to prevent future issues

4. Door Issues

Symptoms:
- Difficulty opening or closing
- Air leaks around edges
- Foggy glass

Troubleshooting steps:
a) Check alignment:
- Ensure hinges are tight and aligned
- Verify frame is square

b) Inspect seals:
- Look for wear or damage on weather stripping
- Check for gaps when door is closed

c) Examine glass (if applicable):
 - Look for cracks or seal failure
 - Clean thoroughly to rule out residue issues

Solution:
 - Adjust or replace hinges
 - Install new weather stripping
 - Replace damaged glass or entire door if necessary

5. Electrical Problems

Symptoms:
 - Controls not responding
 - Intermittent power issues
 - Tripping circuit breakers

Troubleshooting steps:
a) Inspect wiring:
 - Look for signs of heat damage or corrosion
 - Ensure all connections are tight

b) Check control panel:
 - Look for visible damage
 - Verify proper programming

c) Test individual components:
 - Use a multimeter to check continuity
 - Isolate problem areas

Solution:
 - Replace damaged wiring or components
 - Update control panel if outdated
 - Consult a licensed electrician for complex issues

6. Odor Problems

Symptoms:
- Musty smell
- Chemical odor
- Persistent unpleasant scents

Troubleshooting steps:
a) Identify source:
  - Check for mold or mildew
  - Assess recent chemical use
  - Look for trapped moisture

b) Clean thoroughly:
  - Use sauna-safe cleaning products
  - Pay special attention to benches and floors

c) Improve air circulation:
  - Ensure proper ventilation
  - Leave door open between uses

Solution:
- Deep clean and disinfect all surfaces
- Use natural odor absorbers (e.g., charcoal, vinegar)
- Replace any moldy wood or materials

7. Heater Stones Issues

Symptoms:
- Poor heat distribution
- Unusual noises when water is added
- Crumbling stones

Troubleshooting steps:
a) Inspect stones:
- Look for cracks or disintegration
- Check for proper arrangement

b) Clean stones:
- Remove and wash thoroughly
- Discard any damaged stones

c) Verify stone type:
- Ensure stones are suitable for sauna use
- Check if softer stones need more frequent replacement

Solution:
- Rearrange stones for better air flow
- Replace damaged or unsuitable stones
- Use only high-quality sauna stones

General Troubleshooting Tips:

1. Keep a maintenance log:
- Track issues and solutions
- Note regular maintenance tasks

2. Use high-quality materials:
- Invest in sauna-specific products
- Avoid cheap substitutes

3. Regular inspections:
- Perform weekly visual checks
- Address small issues before they become big problems

4. Educate users:
- Provide clear usage instructions
- Encourage reporting of any unusual observations

5. Have a professional assessment:
- Schedule annual inspections by a sauna expert
- Get professional help for complex issues

Remember, many sauna issues can be prevented with proper maintenance and care. When in doubt, always consult with a professional to ensure safe and effective troubleshooting, especially for electrical or structural concerns.

Are you feeling more confident about handling common sauna issues? Which problem-solving skills do you think will be most valuable in maintaining your sauna? Remember, being proactive and addressing small issues promptly can save you time, money, and ensure a consistently enjoyable sauna experience.

Extending the Lifespan of Your Sauna

A well-maintained sauna can provide years of relaxation and health benefits. Let's explore comprehensive strategies to extend the lifespan of your sauna, ensuring it remains a valuable part of your wellness routine for years to come.

Step 1: Establish a Regular Maintenance Schedule

1. Daily tasks:
   - Wipe down benches and walls after each use
   - Leave the door open to air out the sauna

2. Weekly tasks:
   - Deep clean all surfaces
   - Check for any signs of wear or damage

3. Monthly tasks:
   - Inspect electrical components and connections
   - Clean and rearrange heater stones

4. Seasonal tasks:
   - Check and clean ventilation systems
   - Perform a thorough inspection of all sauna components

Step 2: Proper Ventilation

1. Ensure adequate airflow:
   - Keep vents clear of obstructions
   - Consider installing a small fan for improved circulation

2. Manage humidity:
   - Use a hygrometer to monitor humidity levels
   - Aim for 30-40% humidity when the sauna is not in use

3. Allow drying time:
- Keep the door open for at least 30 minutes after each use
- Run ventilation fans to expedite drying

Step 3: Wood Care and Preservation

1. Regular cleaning:
- Use only mild, sauna-safe cleaning products
- Avoid harsh chemicals that can damage wood

2. Sanding:
- Lightly sand benches and walls annually to remove splinters
- Always sand in the direction of the wood grain

3. Apply wood treatment:
- Use food-grade mineral oil or specialized sauna wood treatment
- Reapply every 6-12 months, focusing on high-use areas

Step 4: Protect Against Moisture Damage

1. Maintain proper waterproofing:
- Regularly inspect and reseal grout lines in tiled areas
- Check and replace caulking around joints and seams

2. Use proper sauna etiquette:
- Sit on towels to absorb sweat
- Avoid bringing excess water into the sauna

3. Address leaks immediately:
- Regularly check for signs of water intrusion
- Repair any leaks promptly to prevent wood rot or mold

Step 5: Electrical System Maintenance

1. Regular inspections:
   - Check wiring for signs of wear or heat damage
   - Ensure all connections are tight and corrosion-free

2. Keep controls clean and dry:
   - Use a slightly damp cloth to clean control panels
   - Protect controls from excessive moisture

3. Professional check-ups:
   - Have a licensed electrician inspect the system annually

Step 6: Heater and Stone Maintenance

1. Clean heater regularly:
   - Remove dust and debris from heater surfaces
   - Check for any signs of damage or corrosion

2. Maintain sauna stones:
   - Rearrange stones monthly for even heat distribution
   - Replace any cracked or crumbling stones
   - Completely replace stones annually

Step 7: Door and Window Care

1. Maintain seals:
   - Check weatherstripping regularly and replace if worn
   - Lubricate hinges to ensure smooth operation

2. Clean glass:
   - Use a non-abrasive, sauna-safe glass cleaner
   - Check for any cracks or seal failures

Step 8: Structural Integrity

1. Regular inspections:
   - Check for any loose boards or structural weaknesses
   - Pay attention to bench supports and ceiling integrity

2. Address issues promptly:
   - Tighten any loose screws or nails
   - Replace damaged boards or supports immediately

Step 9: Water Quality (for steam saunas)

1. Use appropriate water:
   - If possible, use distilled or softened water to prevent mineral buildup
   - Clean and descale steam generators regularly

2. Maintain proper drainage:
   - Ensure floor drains are clear and functioning
   - Clean drain covers regularly to prevent clogs

Step 10: User Education

1. Create clear guidelines:
   - Post proper usage instructions in the sauna area
   - Educate all users on sauna etiquette and care

2. Encourage reporting:
   - Ask users to report any unusual sounds, smells, or visible issues

Troubleshooting Tips:

1. Persistent Odors:
- Deep clean with a vinegar solution
- Place a bowl of baking soda in the sauna when not in use
- Check for hidden mold or mildew

2. Uneven Heating:
- Rearrange sauna stones for better air flow
- Check for obstructions around the heater
- Verify that temperature sensors are correctly placed

3. Door Not Sealing Properly:
- Adjust hinges and latch mechanism
- Replace weatherstripping
- Check for warping and replace door if necessary

4. Benches Feeling Rough:
- Sand affected areas with fine-grit sandpaper
- Apply food-grade mineral oil after sanding
- Consider using bench covers for added protection

5. Heater Not Reaching Temperature:
- Check electrical connections
- Verify thermostat accuracy
- Consider upgrading to a more powerful heater if sauna is oversized

6. Excessive Moisture (in dry saunas):
- Improve ventilation
- Check for leaks in the structure
- Use a dehumidifier between sauna sessions

7. Control Panel Malfunctions:
- Check for loose connections
- Protect from excessive moisture
- Consider updating to a newer, more reliable model

8. Wood Discoloration:
- Clean with a mild wood cleaner
- Sand lightly and apply UV-resistant wood treatment
- Consider using bench covers in high-use areas

Remember, the key to extending your sauna's lifespan is consistent care and prompt attention to any issues that arise. By following these steps and addressing problems early, you can ensure that your sauna remains a safe, hygienic, and enjoyable part of your wellness routine for many years.

Are you feeling more confident about maintaining your sauna for the long term? Which aspects of sauna care do you think will be most crucial for extending its lifespan? Remember, a well-maintained sauna not only lasts longer but also provides a more enjoyable and beneficial experience each time you use it.

Chapter 9
Sauna Safety and Best Practices

Temperature and Duration Guidelines

Ensuring safe and effective sauna use is crucial for maximizing health benefits while minimizing risks. Let's explore the important temperature and duration guidelines for different types of saunas, along with best practices for a safe and enjoyable experience.

Temperature Guidelines:

1. Traditional Dry Sauna:
   - Optimal temperature range: 150°F to 195°F (65°C to 90°C)
   - Beginners: Start at lower temperatures, around 150°F to 165°F (65°C to 74°C)
   - Experienced users: May prefer higher temperatures, up to 195°F (90°C)

2. Steam Sauna:
   - Optimal temperature range: 110°F to 120°F (43°C to 49°C)
   - Humidity levels: 100% (fully saturated air)

3. Infrared Sauna:
   - Optimal temperature range: 120°F to 140°F (49°C to 60°C)
   - Lower temperatures due to direct heat penetration

Duration Guidelines:

1. General recommendation:
- 10 to 20 minutes per session for most users

2. Beginners:
- Start with 5 to 10 minutes
- Gradually increase duration over several sessions

3. Experienced users:
- May extend to 20 to 30 minutes, depending on personal tolerance and health status

4. Multiple sessions:
- Allow 20 to 30 minutes of cooling between sessions
- Limit to 2-3 sessions per day

Step-by-Step Sauna Session Guide:

1. Preparation:
- Hydrate well before entering the sauna
- Remove jewelry and metal objects
- Take a quick shower to remove oils and lotions

2. Entering the Sauna:
- Enter slowly, allowing your body to adjust
- Start on the lower bench if using a multi-level sauna

3. During the Session:
- Sit or recline comfortably
- Practice deep, relaxed breathing
- Listen to your body - exit if you feel uncomfortable

4. Cooling Down:
- Exit the sauna slowly to prevent dizziness
- Take a cool shower or use a cold plunge pool
- Rehydrate with water or electrolyte drinks

5. Rest Period:
- Rest for at least 10 minutes before considering another session
- Continue to hydrate

Best Practices:

1. Hydration:
- Drink plenty of water before, during (if comfortable), and after sauna use
- Aim for at least 8 ounces of water per 15-minute session

2. Clothing:
- Wear loose, breathable clothing or a towel
- Avoid synthetic materials that can trap heat

3. Monitoring:
- Use a reliable thermometer and timer in the sauna
- Pay attention to how you feel, not just the clock

4. Gradual Acclimation:
- Increase temperature and duration gradually over time
- Allow your body to adapt to sauna use

5. Sauna Hygiene:
- Sit on a towel to absorb sweat
- Shower before and after sauna use

6. Avoid Alcohol:
- Never use a sauna while under the influence of alcohol
   - Alcohol can increase the risk of dehydration and overheating

Troubleshooting Tips:

1. Feeling Lightheaded or Dizzy:
- Exit the sauna immediately
- Sit or lie down in a cool area
- Hydrate and cool your body gradually

2. Difficulty Breathing:
- Lower the temperature or humidity
- Open the sauna door slightly for fresh air
- Consider using a dry sauna instead of a steam sauna

3. Skin Irritation:
- Reduce session duration
- Ensure you're showering before sauna use
- Consider using a sauna suit to protect skin

4. Excessive Sweating:
- This is normal, but ensure you're rehydrating adequately
- Use electrolyte drinks to replace lost minerals

5. Not Sweating Enough:
- Increase the temperature slightly
- Ensure proper hydration before sauna use
- Consider light exercise before sauna to promote sweating

6. Feeling Too Hot:
- Move to a lower bench in multi-level saunas
- Reduce session duration
- Exit and cool down if discomfort persists

7. Sauna Not Reaching Desired Temperature:
- Check thermometer placement
- Ensure door is properly closed
- Verify heater is functioning correctly

8. Cold Spots in Sauna:
- Check for air leaks around doors or windows
- Ensure proper insulation
- Consider repositioning heater or adding heat reflectors

Remember, while these guidelines are general, individual tolerances can vary. Always consult with a healthcare provider before starting a sauna regimen, especially if you have any pre-existing health conditions or are pregnant.

Are you feeling more confident about using your sauna safely and effectively? Which temperature and duration guidelines seem most appropriate for your personal sauna goals? Remember, the key to a beneficial sauna experience is listening to your body and gradually building your tolerance over time.

Hydration and Health Precautions

Proper hydration and health precautions are crucial for a safe and beneficial sauna experience. Let's explore these important aspects in detail to ensure you're using your sauna responsibly and effectively.

Hydration Guidelines:

1. Pre-Sauna Hydration:
   - Drink 16-24 oz (470-710 ml) of water 1-2 hours before your sauna session
   - Avoid caffeine and alcohol, which can contribute to dehydration

2. During Sauna Use:
   - Keep a water bottle nearby
   - Sip water as needed, especially if your session exceeds 15 minutes

3. Post-Sauna Hydration:
   - Drink at least 16-24 oz (470-710 ml) of water after your session
   - Consider electrolyte-rich drinks for longer or more intense sessions

4. Ongoing Hydration:
   - Maintain good hydration throughout the day
   - Aim for pale yellow urine as an indicator of proper hydration

Step-by-Step Hydration Process:

1. Prepare:
- Fill a large water bottle before your sauna session
- Consider adding electrolytes or a pinch of salt for mineral balance

2. Pre-Sauna:
- Drink 8-12 oz (240-350 ml) of water 30 minutes before entering the sauna

3. During Sauna:
- Take small sips of water every 5-10 minutes if comfortable
- Listen to your body's thirst signals

4. Post-Sauna:
- Drink 8-12 oz (240-350 ml) immediately after exiting
- Continue to hydrate over the next hour, aiming for at least 16 oz (470 ml)

5. Monitor:
- Check urine color to ensure you're adequately hydrated
- Weigh yourself before and after to gauge fluid loss if curious

Health Precautions:

1. Consult Your Doctor:
- Always check with a healthcare provider before starting sauna use, especially if you have:
  • Heart conditions
  • High or low blood pressure
  • Pregnancy
  • Chronic health conditions

2. Acclimate Gradually:
- Start with shorter sessions (5-10 minutes) at lower temperatures
- Gradually increase time and temperature over several weeks

3. Listen to Your Body:
- Exit the sauna if you feel dizzy, nauseous, or excessively uncomfortable
- Don't push beyond your comfort level

4. Avoid Sauna Use If:
- You're ill or have a fever
- You've recently consumed alcohol
- You're severely sleep-deprived

5. Cooling Down:
- Exit the sauna slowly to prevent dizziness
- Cool down gradually; avoid extreme temperature changes

6. Rest and Recover:
- Allow your body time to recover between sauna sessions
- Don't use the sauna multiple times a day until you're well-acclimated

7. Hygiene Practices:
- Shower before entering the sauna to remove oils and lotions
- Sit on a towel to maintain cleanliness

8. Jewelry and Contact Lenses:
- Remove metal jewelry to prevent burns
- Consider removing contact lenses to avoid eye irritation

Troubleshooting Tips:

1. Feeling Dehydrated:
- Increase water intake before and after sessions
- Reduce sauna duration or frequency
- Add electrolytes to your water

2. Headache After Sauna Use:
- This often indicates dehydration
- Increase pre- and post-sauna water intake
- Consider adding a pinch of salt to your water for better absorption

3. Dizziness or Lightheadedness:
- Exit the sauna immediately
- Lie down in a cool area with feet elevated
- Sip water slowly and cool down gradually

4. Skin Irritation or Excessive Dryness:
- Reduce sauna duration or frequency
- Apply a moisturizer after showering post-sauna
- Consider using a humidifier in dry saunas

5. Difficulty Breathing:
- This could indicate overheating or dehydration
- Exit the sauna and cool down
- If persistent, consult a healthcare provider

6. Muscle Cramps:
- Often a sign of electrolyte imbalance
- Increase intake of electrolyte-rich fluids
- Consider adding magnesium supplements (consult your doctor first)

7. Feeling Overly Fatigued After Sauna:
- Reduce session duration
- Ensure proper hydration and electrolyte balance
- Allow more recovery time between sessions

8. Rapid Heartbeat:
- Exit the sauna and cool down gradually
- If persistent or concerning, seek medical attention
- Consult your doctor about safe sauna use if you have heart conditions

9. Nausea:
- This can indicate overheating or dehydration
- Exit the sauna, cool down, and sip water slowly
- Reduce session duration in future uses

10. Persistent Thirst Despite Drinking Water:
- Increase electrolyte intake
- Consider using an oral rehydration solution
- Consult a healthcare provider if thirst persists

Remember, proper hydration and adherence to health precautions are key to enjoying the benefits of sauna use while minimizing risks. Always prioritize safety and listen to your body's signals. If you have any concerns or experience persistent issues, don't hesitate to consult with a healthcare professional.

Are you feeling more confident about maintaining proper hydration and following health precautions during your sauna sessions? Which aspects of hydration and health safety do you think will be most important for your personal sauna routine? Remember, a safe and enjoyable sauna experience starts with taking care of your body's basic needs, especially hydration.

Fire Safety and Electrical Considerations

Ensuring fire safety and proper electrical setup in your sauna is crucial for preventing accidents and maintaining a safe environment. Let's explore these important aspects in detail.

Fire Safety Measures:

Step 1: Proper Installation and Clearances

1. Heater placement:
   - Follow manufacturer's guidelines for minimum clearances
   - Typically, maintain at least 4 inches from walls

2. Protective barriers:
   - Install a non-combustible barrier around the heater
   - Use heat-resistant materials like stone or tile

Step 2: Fire-Resistant Materials

1. Choose appropriate wood:
   - Use naturally fire-resistant woods like cedar or hemlock
   - Avoid resinous woods like pine

2. Treat wood surfaces:
   - Apply fire-retardant treatments to wood surfaces
   - Reapply as recommended by the product manufacturer

Step 3: Ventilation

1. Ensure proper airflow:
   - Install vents as per sauna design specifications
   - Keep vents clear of obstructions

2. Avoid overheating:
- Use a thermometer to monitor temperature
- Don't exceed manufacturer's recommended maximum temperature

Step 4: Fire Detection and Suppression

1. Install smoke detectors:
- Place outside the sauna room
- Test regularly and replace batteries as needed

2. Fire extinguisher:
- Keep a suitable fire extinguisher near the sauna
- Ensure all users know its location and how to use it

Electrical Safety Considerations:

Step 1: Professional Installation

1. Hire a licensed electrician:
- Ensure all wiring meets local codes
- Obtain necessary permits and inspections

2. Use appropriate materials:
- Use heat-resistant wiring suitable for sauna conditions
- Install proper conduit for all electrical runs

Step 2: Proper Grounding and Circuit Protection

1. GFCI protection:
- Install GFCI (Ground Fault Circuit Interrupter) on sauna circuits
- Test GFCI monthly

2. Dedicated circuit:
- Use a dedicated circuit for the sauna
- Ensure proper amperage based on heater requirements

Step 3: Heater and Control Installation

1. Follow manufacturer's instructions:
- Adhere strictly to installation guidelines
- Use only approved components

2. Secure mounting:
- Ensure heater is securely fastened
- Mount control panel as per manufacturer's specifications

Step 4: Regular Inspections and Maintenance

1. Visual inspections:
- Check for signs of wear, damage, or corrosion regularly
- Look for loose connections or frayed wires

2. Professional check-ups:
- Schedule annual inspections by a licensed electrician
- Address any issues promptly

Troubleshooting Tips:

1. Heater Not Turning On:
- Check circuit breaker and reset if tripped
- Verify all connections are secure
- Test GFCI and reset if necessary

2. Inconsistent Heating:
- Check for loose wires or connections
- Ensure temperature sensor is properly placed and functioning

- Verify heater elements are all working

3. Burning Smell:
   - Turn off sauna immediately
   - Inspect for any visible signs of overheating or burning
   - Have a professional check wiring and components

4. Flickering Lights:
   - Check for loose bulbs or fixtures
   - Verify proper wattage is being used
   - Inspect wiring for any damage

5. Control Panel Malfunctions:
   - Ensure connections are secure
   - Check for any moisture damage
   - Consider replacing if issues persist

6. Tripping Circuit Breaker:
   - Verify circuit is properly sized for the heater
   - Check for any short circuits in the wiring
   - Consider upgrading the electrical service if necessary

7. Overheating:
   - Check temperature sensor placement and function
   - Verify heater is not exceeding its rated capacity
   - Ensure proper ventilation in the sauna room

8. Electrical Shocks:
   - Turn off power immediately
   - Check for any water leaks or high humidity issues
   - Have a professional inspect the entire electrical system

9. Visible Sparks or Arcing:
- Turn off power immediately
- Do not use the sauna until inspected by a professional
- Check for any damaged wires or loose connections

10. Strange Noises from Heater:
- Turn off and allow to cool
- Inspect for any foreign objects in the heater
- Check for loose components or connections

Remember, when it comes to fire safety and electrical considerations, prevention is key. Regular maintenance, proper installation, and prompt attention to any issues are crucial. Never attempt to repair electrical components yourself unless you are a qualified professional. Safety should always be the top priority in sauna operation and maintenance.

Are you feeling more confident about maintaining fire safety and electrical integrity in your sauna? Which aspects of these safety measures do you think will require the most attention in your specific sauna setup? Remember, a safe sauna is a enjoyable sauna, and these precautions ensure you can relax without worry during your sessions.

Child and Pet Safety Measures

Ensuring the safety of children and pets around your sauna is crucial. Let's explore comprehensive measures to create a safe environment and prevent accidents.

Step 1: Secure the Sauna Area

1. Install a lockable door:
    - Use a childproof lock or a high latch out of children's reach
    - Ensure the lock is easy for adults to operate in case of emergency

2. Create a safety barrier:
    - Install a fence or gate around outdoor saunas
    - Use a room divider or screen for indoor saunas

3. Post clear warning signs:
    - Use large, easy-to-read text
    - Include pictograms for non-readers

Step 2: Implement Temperature Controls

1. Install a tamper-proof thermostat:
    - Place controls out of children's reach
    - Use a lockable cover for the control panel

2. Set temperature limits:
    - Program maximum temperature settings
    - Consider a dual-control system requiring two actions to change settings

Step 3: Establish Clear Rules and Education

1. Create age-appropriate rules:
- No unsupervised sauna use for children
- Set time limits for supervised use (e.g., 5-10 minutes max for older children)

2. Educate family members:
- Explain sauna risks and safety measures
- Teach proper sauna etiquette and usage

3. Demonstrate emergency procedures:
- Show how to exit quickly
- Teach how to lower temperature or turn off the sauna

Step 4: Pet-Specific Precautions

1. Keep pets out:
- Always close the sauna door
- Use pet gates to block access to the sauna area

2. Create a pet-safe zone:
- Provide a comfortable area away from the sauna for pets
- Ensure this area has water and proper ventilation

Step 5: Implement Additional Safety Features

1. Install a timer:
- Use an automatic shut-off feature
- Set maximum runtime limits

2. Add proper lighting:
- Ensure the sauna and surrounding area are well-lit
- Consider motion-activated lights for better visibility

3. Maintain proper ventilation:
- Ensure vents are not blocked
- Install a fan if necessary for air circulation

Step 6: Regular Safety Checks

1. Conduct monthly inspections:
- Check locks, latches, and barriers
- Test temperature controls and timers

2. Maintain cleanliness:
- Keep the sauna and surrounding area free of trip hazards
- Regularly clean to prevent slippery surfaces

Step 7: Emergency Preparedness

1. Install a communication device:
- Keep a waterproof phone or intercom system nearby
- Ensure it's easily accessible in case of emergency

2. First aid kit:
- Keep a well-stocked first aid kit near the sauna
- Include burn treatment supplies

Troubleshooting Tips:

1. Child Accidentally Enters Sauna:
- Immediately turn off the sauna
- Remove the child and assess for any signs of distress
- Provide cool water and monitor for symptoms of overheating

2. Pet Shows Interest in Sauna:
- Reinforce barriers and locks
- Use deterrent sprays or mats around the sauna area
- Increase supervision during sauna use

3. Lock Malfunction:
- Implement a backup locking system (e.g., portable lock)
- Schedule immediate repair or replacement
- Increase supervision until fixed

4. Children Tampering with Controls:
- Install a more secure control panel cover
- Move controls to a higher, less accessible location
- Implement stricter supervision rules

5. Persistent Attempts to Enter:
- Re-educate about sauna dangers
- Consider additional deterrents (e.g., alarms)
- Evaluate if sauna location needs to be changed

6. Overheating Incidents:
- Double-check temperature limit settings
- Install an additional temperature sensor
- Increase frequency of safety checks

7. Pets Overheating in Nearby Areas:
- Improve ventilation in surrounding spaces
- Create a cooler, designated pet area further from the sauna
- Consider restricting pet access during sauna use

8. Children Sharing Unsafe Information:
- Have a family meeting to reinforce safety rules
- Create a "sauna safety pledge" for children to sign
- Involve children in age-appropriate safety checks

9. Accidental Activation:
- Install a master power switch out of reach
- Use a cover for the start button
- Implement a two-step activation process

10. Curiosity About Sauna Stones:
- Install a secure guard around the heater
- Educate about the dangers of touching hot stones
- Supervise closely during any educational demonstrations

Remember, the key to child and pet safety around saunas is a combination of physical barriers, clear rules, education, and constant vigilance. Regular reinforcement of safety measures and open communication about sauna risks are essential for maintaining a safe environment.

Are you feeling more prepared to ensure child and pet safety around your sauna? Which safety measures do you think will be most crucial for your specific situation? Remember, creating a safe environment allows everyone in the household to enjoy the benefits of the sauna without worry.

Chapter 10
Enhancing Your Sauna Experience
Traditional Sauna Rituals

Traditional sauna rituals can transform your sauna sessions from simple heat therapy into a rich, cultural experience. Let's explore some time-honored practices and how to incorporate them into your routine.

Step 1: Preparation

1. Hydrate well:
   - Drink 16-24 oz of water 1-2 hours before your session

2. Gather supplies:
   - Towels (at least 2-3)
   - Water bottle
   - Natural exfoliating brush or vihta/vasta (birch branches)
   - Optional: Essential oils, honey, or salt for skin treatments

3. Set the mood:
   - Light candles or adjust lighting for a relaxing ambiance
   - Prepare cool water or ice for contrast therapy

Step 2: The Pre-Sauna Ritual

1. Take a warm shower:
   - Cleanse your body without using soap
   - This helps open pores and begin the warming process

2. Dry off completely:
   - Excess moisture can interfere with sweating

3. Enter the sauna nude or wrapped in a towel:
- This allows for maximum skin exposure to heat

Step 3: The First Round

1. Start with lower temperatures:
- Begin at around 150-160°F (65-70°C)
- Sit on the lower bench if you're sensitive to heat

2. Practice deep, relaxed breathing:
- Inhale through your nose, exhale through your mouth
- Focus on the sensation of heat on your skin

3. Stay in for 10-15 minutes:
- Listen to your body and exit if you feel uncomfortable

Step 4: The Cool-Down

1. Exit the sauna slowly:
- Move carefully to prevent dizziness

2. Take a cool shower or plunge:
- Start with lukewarm water and gradually decrease temperature
- For the brave, a quick dip in a cold pool or snow roll (if available)

3. Rest and hydrate:
- Sit or lie down in a cool area
- Drink water or a natural electrolyte drink

Step 5: The Second Round

1. Re-enter the sauna:
- The temperature may feel more intense now

2. Löyly ritual (adding water to stones):
- Slowly ladle water onto the hot stones
- Feel the burst of humid heat (steam)
- Use essential oils like eucalyptus for aromatherapy if desired

3. Use a vihta/vasta (optional):
- Gently brush your skin with birch branches
- This stimulates circulation and exfoliates

Step 6: Final Cool-Down and Recovery

1. Take a final cool shower or plunge
2. Dry off and rest:
- Allow your body temperature to normalize
- This can take 10-20 minutes

3. Rehydrate and nourish:
- Drink plenty of water
- Consider a light, healthy snack

Step 7: Skin Care

1. Apply a natural moisturizer:
- Use coconut oil or aloe vera to soothe and hydrate skin

2. Facial treatment (optional):
- Apply a cool, damp cloth to your face
- Use a gentle facial moisturizer

Enhancing the Experience:

1. Soundscape:
- Play soft, natural sounds or traditional Finnish music

2. Aromatherapy:
- Use sauna-safe essential oils like eucalyptus or lavender

3. Meditation or mindfulness:
- Practice guided meditation or focus on your breath

4. Social aspect:
- Invite friends or family for a communal experience
- Engage in quiet conversation (traditional in many sauna cultures)

Troubleshooting Tips:

1. Feeling Overwhelmed by Heat:
- Move to a lower bench
- Open the door slightly for fresh air
- Exit and cool down if discomfort persists

2. Dry Skin or Irritation:
- Reduce session duration
- Increase water intake
- Apply natural oils like coconut oil before and after sauna use

3. Dizziness During Cool-Down:
- Transition temperatures more gradually
- Sit down immediately if feeling faint
- Ensure proper hydration before and after sauna use

4. Difficulty with Löyly (Steam):
- Start with small amounts of water on stones
- Keep face away from direct steam
- Use cooler water for a gentler effect

5. Vihta/Vasta Causing Skin Irritation:
- Soak branches in water longer before use
- Use gentler strokes
- Consider skipping this step if sensitivity persists

6. Overheating:
- Shorten sauna sessions
- Increase cool-down periods
- Monitor your body's signals closely

7. Dehydration Symptoms:
- Increase water intake before and after sessions
- Add electrolytes to your water
- Reduce sauna frequency until hydration improves

8. Difficulty Relaxing:
- Practice deep breathing exercises
- Try guided meditation apps
- Adjust lighting or sound for a more soothing atmosphere

Remember, traditional sauna rituals are about listening to your body and finding what works best for you. It's okay to modify practices to suit your comfort level and health needs. The goal is to create a relaxing, rejuvenating experience that you look forward to regularly.

Are you excited to incorporate some of these traditional rituals into your sauna routine? Which aspects of the ritual do you think will enhance your experience the most? Remember, the beauty of sauna traditions lies in their adaptability – feel free to create your own personal ritual that brings you the most joy and relaxation.

Combining Hot and Cold Therapies

Alternating between hot and cold temperatures, known as contrast therapy, can enhance the benefits of your sauna experience. This practice can improve circulation, reduce muscle soreness, and boost overall well-being. Let's explore how to safely and effectively combine hot and cold therapies.

Step 1: Preparation

1. Set up your hot and cold stations:
   - Sauna (hot therapy)
   - Cold shower, plunge pool, or ice bath (cold therapy)

2. Gather supplies:
   - Multiple towels
   - Robe or wrap
   - Non-slip footwear
   - Water bottle
   - Timer or watch

3. Check temperatures:
   - Sauna: 170-190°F (76-88°C)
   - Cold therapy: 50-59°F (10-15°C) for beginners, colder for experienced users

Step 2: Begin with Heat

1. Enter the sauna:
   - Start with 10-15 minutes
   - Focus on relaxed, deep breathing

2. Allow your body to warm up:
   - Sweat should begin within a few minutes
   - Stay mindful of how you feel

Step 3: Transition to Cold

1. Exit the sauna slowly:
- Move carefully to prevent dizziness

2. Quickly move to your cold therapy station:
- Submerge in a cold plunge pool, or
- Take a cold shower, starting with feet and moving upward

3. Cold exposure time:
- Beginners: 30 seconds to 1 minute
- Experienced users: Up to 3 minutes

4. Focus on controlled breathing:
- Take slow, deep breaths to manage the cold shock

Step 4: Return to Heat

1. Re-enter the sauna:
- Your body will warm up faster this time

2. Stay for 5-10 minutes:
- Notice the tingling sensation as your body reheats

Step 5: Repeat the Cycle

1. Alternate between hot and cold:
- Aim for 3-4 cycles total

2. Gradually increase cold exposure time:
  - Add 15-30 seconds each session as you become accustomed

3. Listen to your body:
- Stop if you feel uncomfortable or dizzy

Step 6: Final Cool Down

1. End with a cold session:
- This helps close pores and invigorates the body

2. Take a lukewarm shower:
- Rinse off sweat and any impurities

Step 7: Recovery

1. Rest and rehydrate:
- Sit or lie down in a neutral temperature area
- Drink water or an electrolyte beverage

2. Allow your body to return to normal temperature:
- This usually takes 15-20 minutes

Enhancing the Experience:

1. Use aromatherapy:
- Add eucalyptus or peppermint oil to cold water for an invigorating scent

2. Practice mindfulness:
- Focus on bodily sensations during temperature changes

3. Incorporate gentle stretching:
- Do light stretches between hot and cold sessions

4. Try contrast showers:
- Alternate hot and cold water if a plunge pool isn't available

Troubleshooting Tips:

1. Cold Shock is Too Intense:
- Start with slightly warmer water (around 60°F or 15°C)
- Gradually lower temperature over multiple sessions
- Begin with shorter cold exposures (15-30 seconds)

2. Dizziness During Transitions:
- Move more slowly between hot and cold
- Take a seated position during cold therapy
- Ensure proper hydration before and during the session

3. Skin Irritation or Redness:
- Reduce time in both hot and cold therapies
- Apply a gentle, natural moisturizer after the session
- Consult a dermatologist if persistent

4. Muscle Cramping:
- Increase hydration and electrolyte intake
- Perform gentle stretches between cycles
- Reduce intensity or duration of contrast therapy

5. Difficulty Warming Up After Cold:
- Increase time in the sauna
- Use a warm towel or robe between transitions
- Consider adding an intermediate temperature step

6. Feeling Overly Fatigued:
- Reduce the number of cycles
- Increase rest time between sessions
- Ensure adequate sleep and nutrition before contrast therapy

7. Rapid Heart Rate:
- Extend rest periods between hot and cold
- Practice deep, controlled breathing
- Consult a doctor if this persists, especially if you have heart conditions

8. Cold Water Fear:
- Start with cool rather than cold water
- Use a shower instead of full immersion
- Practice gradual exposure techniques

9. Headache After Session:
- Ensure proper hydration before, during, and after
- Reduce intensity or duration of sessions
- Apply a cool compress to the forehead and neck

10. Difficulty Sleeping After Evening Sessions:
- Schedule contrast therapy earlier in the day
- End with a longer cool-down period
- Practice relaxation techniques post-session

Remember, while contrast therapy can offer numerous benefits, it's important to introduce it gradually and listen to your body. Start conservatively and build up your tolerance over time. If you have any health concerns or conditions, always consult with a healthcare professional before starting contrast therapy.

Are you excited to try combining hot and cold therapies in your sauna routine? Which aspect of this practice do you think will be most challenging or beneficial for you? Remember, the key to a successful contrast therapy experience is finding the right balance that invigorates and refreshes you without causing undue stress on your body.

Meditation and Relaxation Techniques

Incorporating meditation and relaxation techniques into your sauna sessions can significantly enhance the mental and emotional benefits of your experience. Let's explore how to effectively combine these practices with your sauna routine.

Step 1: Prepare Your Sauna Environment

1. Set the mood:
   - Dim the lights or use candles (safely placed)
   - Ensure proper ventilation
   - Set a comfortable temperature (around 150-170°F or 65-75°C)

2. Gather supplies:
   - Comfortable seating (towel or mat)
   - Water bottle
   - Optional: aromatherapy oils, singing bowl, or soft music

Step 2: Begin with a Breathing Exercise

1. Seated Comfortably:
   - Sit on the lower bench if you're new to sauna meditation

2. Practice 4-7-8 Breathing:
   - Inhale through your nose for 4 counts
   - Hold your breath for 7 counts
   - Exhale slowly through your mouth for 8 counts
   - Repeat 3-4 times

Step 3: Body Scan Meditation

1. Start at your toes:
   - Focus your attention on your toes, noticing any sensations

2. Slowly move up your body:
- Progress through feet, legs, torso, arms, and head
- Spend 15-30 seconds on each body part

3. Release tension:
- As you focus on each area, consciously relax those muscles

Step 4: Mindfulness Meditation

1. Focus on physical sensations:
- Notice the heat on your skin
- Pay attention to your breath
- Observe sweat forming and trickling

2. Acknowledge thoughts without judgment:
   - If your mind wanders, gently bring focus back to sensations

3. Use a mantra (optional):
- Repeat a calming phrase like "I am relaxed and at peace"

Step 5: Visualization Exercise

1. Imagine a peaceful scene:
- Picture a serene beach, forest, or mountain top

2. Engage all senses:
- What do you see, hear, smell, and feel in this place?

3. Immerse yourself:
- Spend 5-10 minutes in this imaginary location

Step 6: Gratitude Practice

1. Reflect on positive aspects of your life:
- Think of 3-5 things you're grateful for

2. Feel the gratitude:
- Allow feelings of appreciation to fill your body

3. Send positive thoughts:
- Direct well-wishes to yourself and others

Step 7: Progressive Muscle Relaxation

1. Start with your feet:
- Tense the muscles in your feet for 5 seconds
- Release and relax for 10 seconds

2. Move up your body:
- Repeat the process with each muscle group
- End with facial muscles

Step 8: Closing Your Practice

1. Take a few deep breaths:
- Slowly bring your awareness back to the present

2. Express gratitude:
- Thank yourself for taking this time for self-care

3. Slowly open your eyes:
- Take a moment to adjust before moving

Enhancing Your Practice:

1. Use aromatherapy:
   - Add calming scents like lavender or eucalyptus

2. Incorporate sound therapy:
   - Use a singing bowl or play soft, ambient music

3. Try guided meditations:
   - Use apps or recordings designed for sauna use

Troubleshooting Tips:

1. Difficulty Concentrating:
   - Start with shorter sessions (5-10 minutes)
   - Use a focal point like a candle flame or crystal
   - Try counting breaths to anchor your attention

2. Feeling Overwhelmed by Heat:
   - Move to a lower bench
   - Reduce session duration
   - Focus on cool visualizations (e.g., snowy landscapes)

3. Physical Discomfort:
   - Adjust your position
   - Use a rolled towel for back support
   - Practice on a cooler day when starting out

4. Racing Thoughts:
   - Acknowledge thoughts without engaging them
   - Return focus to your breath or body sensations
   - Try a mantra or counting meditation

5. Falling Asleep:
- Sit upright rather than reclining
- Open your eyes slightly
- Practice earlier in the day when you're more alert

6. Time Management Stress:
- Use a gentle timer or alarm
- Practice accepting whatever length of time you have

7. Sweating Distractions:
- Keep a small towel handy
- Incorporate awareness of sweat into your mindfulness practice

8. Dehydration Affecting Focus:
- Hydrate well before your session
- Keep water nearby and sip as needed

9. Difficulty with Visualization:
- Start with simple, familiar scenes
- Focus on one sense at a time
- Use a physical object as a starting point

10. Emotional Releases:
- Allow emotions to surface without judgment
- Practice self-compassion
- Consider journaling after your session

Remember, meditation and relaxation in the sauna is a practice. It may take time to find the techniques that work best for you. Be patient with yourself and approach each session with an open mind. The combination of heat therapy and mindfulness can lead to profound relaxation and stress relief when practiced regularly.

Are you excited to incorporate these meditation and relaxation techniques into your sauna routine? Which method do you think will be most beneficial for your mental and emotional well-being? Remember, the goal is to create a peaceful, rejuvenating experience that leaves you feeling refreshed and centered long after you've left the sauna.

Sauna Socializing and Etiquette

Sauna socializing can be a wonderful way to bond with friends and family while enjoying the health benefits of heat therapy. However, it's important to follow proper etiquette to ensure everyone has a comfortable and enjoyable experience. Let's explore the steps to successful sauna socializing and the etiquette rules to follow.

Step 1: Preparing for Social Sauna Time

1. Invite appropriately:
    - Consider the size of your sauna and invite accordingly
    - Ensure all guests are comfortable with shared sauna use

2. Set expectations:
    - Inform guests about your sauna's traditions and rules
    - Communicate expected duration and any planned activities

3. Prepare the space:
    - Ensure the sauna is clean and well-ventilated
    - Provide enough towels, water, and seating for all guests

Step 2: Welcoming Guests

1. Offer a tour:
    - Show where to change, store belongings, and shower
    - Explain how to use the sauna controls and add water to stones (if applicable)

2. Provide amenities:
    - Offer water or other hydrating drinks
    - Have extra towels and flip-flops available

Step 3: Entering the Sauna

1. Shower before entering:
   - Remind guests to rinse off and enter the sauna clean

2. Proper attire:
   - Follow local customs (nude, towel-wrapped, or swimwear)
   - Ensure all guests are comfortable with the chosen attire

3. Seating arrangement:
   - Higher benches for those who prefer more heat
   - Lower benches for those who prefer less intense heat

Step 4: Sauna Conversation Etiquette

1. Keep volume moderate:
   - Speak in normal tones; avoid shouting

2. Choose light topics:
   - Engage in pleasant, stress-free conversation
   - Avoid controversial or heated debates

3. Respect silence:
   - It's okay to have quiet moments
   - Some may prefer to meditate or relax silently

Step 5: Managing Heat and Steam

1. Ask before adding water to stones:
   - Ensure all guests are comfortable with increased steam

2. Respect heat preferences:
   - Allow guests to move to lower benches if needed
   - Don't pressure others to stay longer than they're comfortable

Step 6: Exiting and Cooling Down

1. Leave quietly:
   - Exit without disrupting others if leaving early

2. Cool down together:
   - Offer cold plunge, shower, or outdoor cool-down options
   - This can be a great time for continued socializing

Step 7: Post-Sauna Socializing

1. Provide a relaxation area:
   - Offer comfortable seating outside the sauna
   - This allows for continued conversation in a cooler environment

2. Offer refreshments:
   - Provide water, herbal tea, or traditional post-sauna snacks

Sauna Etiquette Rules:

1. Respect personal space:
   - Don't spread out more than necessary
   - Ask before adjusting shared items (e.g., thermometer, ladle)

2. Maintain hygiene:
   - Sit on a towel
   - Don't groom or shave in the sauna

3. Be mindful of scents:
   - Avoid strong perfumes or lotions
   - Ask before using aromatherapy oils

4. Honor time limits:
- Don't overstay if others are waiting
- Follow any posted time guidelines

5. Clean up after yourself:
- Wipe down your seating area
- Take all personal items when you leave

Troubleshooting Tips:

1. Conflicting Temperature Preferences:
- Use a tiered seating system (hotter up top, cooler below)
- Take breaks and re-enter together

2. Excessive Talking:
- Gently suggest a quiet period for relaxation
- Lead by example, starting a period of peaceful silence

3. Uncomfortable with Nudity:
- Offer a swimwear option for all
- Create separate sessions for those comfortable with nudity

4. Overheating Guest:
- Encourage them to step out and cool down
- Offer a cool, damp towel

5. Inappropriate Behavior:
- Address issues privately and immediately
- Have clear rules posted to refer to if needed

6. Dehydration Concerns:
- Keep a water station visible and accessible
- Remind guests to hydrate regularly

7. Conflicting Aromatherapy Preferences:
- Use neutral scents or no scents for group sessions
- Offer personal inhalers for individual aromatherapy

8. Time Management Issues:
- Use a visible timer
- Agree on session duration beforehand

9. Personal Space Invasion:
- Arrange seating to create natural spaces between guests
- Politely remind guests about respecting others' space

10. Post-Sauna Chill:
- Provide warm robes or blankets
- Ensure a comfortable temperature in the relaxation area

Remember, successful sauna socializing is about creating a welcoming, respectful environment where everyone can relax and enjoy the experience. By following these etiquette guidelines and being considerate of others, you can foster a positive sauna culture among your friends and family.

Are you looking forward to hosting a social sauna session? Which aspects of sauna etiquette do you think will be most important for your group? Remember, the goal is to create an atmosphere where everyone feels comfortable and can fully enjoy the physical and social benefits of the sauna experience.

Appendices

Sauna Wood Types and Their Properties

Choosing the right wood for your sauna is crucial for durability, aesthetics, and overall experience. Here's a comprehensive guide to the most popular sauna wood types and their properties:

1. Western Red Cedar

Properties:
- Excellent insulation properties
- Natural resistance to decay and insects
- Pleasant, distinctive aroma
- Low heat conductivity
- Minimal shrinkage and warping
- Light weight

Pros:
- Ideal for high-humidity environments
- Long-lasting
- Aesthetically pleasing with rich color variations

Cons:
- Can be expensive
- Some people may be sensitive to its strong aroma

Best for: Traditional saunas, especially in humid climates

2. Hemlock

Properties:
- Light color with straight grain
- Good strength-to-weight ratio

- Low resin content
- Moderate decay resistance

Pros:
- More affordable than cedar
- Minimal scent, good for those sensitive to strong aromas
- Takes stains well for customized appearance

Cons:
- Less naturally resistant to decay than cedar
- May require more maintenance over time

Best for: Budget-conscious sauna builders, indoor saunas

3. Nordic White Spruce

Properties:
- Light color with minimal knots
- Straight grain
- Low resin content
- Moderate durability

Pros:
- Traditional choice in Scandinavian saunas
- Clean, bright appearance
- Affordable option

Cons:
- Less naturally resistant to decay than cedar
- May require more frequent maintenance

Best for: Authentic Finnish-style saunas, bright interior aesthetics

4. Basswood

Properties:
- Very light color, almost white
- Soft texture
- Low heat conductivity
- Minimal scent

Pros:
- Ideal for those with sensitivities to wood aromas
- Stays cool to the touch, good for benches and backrests
- Takes stains well for customization

Cons:
- Less durable than harder woods
- May dent or scratch more easily

Best for: Sauna benches, headrests, and for users with sensitivities

5. Aspen

Properties:
- Light color with subtle grain
- Low heat conductivity
- Minimal tendency to splinter
- Little to no scent

Pros:
- Stays cool to the touch
- Good choice for benches and seating
- Aesthetically pleasing light appearance

Cons:
- Less resistant to decay than cedar
- May require more maintenance in high-humidity environments

Best for: Interior accents, benches, and in combination with other woods

6. Alder

Properties:
- Light color with subtle reddish tinge
- Straight grain
- Moderate hardness
- Low resin content

Pros:
- Affordable option
- Takes stains well for customized appearance
- Minimal scent

Cons:
- Less naturally resistant to decay
- May require more frequent maintenance

Best for: Budget-friendly sauna projects, interior paneling

7. Eucalyptus

Properties:
- Dense and durable
- Natural resistance to decay
- Pleasant, subtle aroma
- Rich color variations

Pros:
- Excellent durability in high-humidity environments
- Unique aesthetic appeal
- Environmentally sustainable (fast-growing)

Cons:
- Can be expensive
- May be harder to source than traditional sauna woods

Best for: Luxury saunas, outdoor saunas in humid climates

When selecting wood for your sauna, consider these factors:
- Local availability and cost
- Personal preferences for aroma and appearance
- The specific area of the sauna (e.g., benches vs. walls)
- Your local climate and humidity levels
- Maintenance requirements and long-term durability

Remember, proper treatment and maintenance of any wood type will significantly extend its lifespan and preserve its properties. Regular cleaning, appropriate oiling, and good ventilation are key to maintaining the beauty and functionality of your sauna wood.

Heater and Stove Comparison Guide

Choosing the right heater or stove for your sauna is crucial for performance, efficiency, and safety. This guide compares different types of sauna heaters to help you make an informed decision.

1. Electric Sauna Heaters

Types:
a) Traditional Electric Heaters
b) Combi Heaters (with steam generation)

Pros:
- Easy to install and use
- Precise temperature control
- Low maintenance
- Clean operation

Cons:
- Require electrical installation
- Ongoing electricity costs
- May lack the traditional sauna ambiance

Best for: Home saunas, urban settings, convenience-oriented users

Power range: 3-18 kW

Key features to consider:
- Built-in controls vs. separate control panel
- Stone capacity
- Heating element quality
- Safety features (overheat protection, timer)

2. Wood-Burning Sauna Stoves

Types:
a) Traditional log-burning stoves
b) Wood-burning stoves with glass doors

Pros:
- Authentic traditional experience
- No electricity required
- Can heat water for steam
- Pleasant wood aroma

Cons:
- Requires regular fuel (wood) and ash cleaning
- More challenging to control temperature
- Needs proper ventilation and chimney

Best for: Traditional sauna enthusiasts, rural settings, off-grid locations

Heat output: Varies, typically sufficient for saunas up to 425 cubic feet

Key features to consider:
- Stone capacity
- Firebox size
- Ease of loading and cleaning
- Safety features (door locks, heat shields)

3. Gas Sauna Heaters

Types:
a) Natural gas heaters
b) Propane heaters

Pros:
- Efficient heating
- Lower operating costs than electric in some areas
- Can provide high heat output

Cons:
- Requires professional installation
- Need for gas line or propane tank
- Potential safety concerns with gas usage

Best for: Large saunas, commercial settings, areas with low gas prices

Power range: 6-18 kW (or higher for commercial use)

Key features to consider:
- Gas type compatibility
- Ventilation requirements
- Safety features (gas leak detection, automatic shut-off)

4. Infrared Sauna Heaters

Types:
a) Ceramic infrared heaters
b) Carbon fiber infrared panels

Pros:
- Lower operating temperature
- Energy efficient
- Quick heat-up time
- No need for preheating

Cons:
- Different experience from traditional saunas
- Limited steam options
- May have less even heat distribution

Best for: Users seeking targeted heat therapy, those sensitive to high temperatures

Power range: 1-2.3 kW per panel

Key features to consider:
- Type of infrared (near, mid, or far)
- Coverage area per panel
- EMF levels

5. Combi Heaters

Combination of traditional electric heater with steam generation capability

Pros:
- Versatile: can function as both dry and steam sauna
- Convenient humidity control
- All-in-one solution

Cons:
- More complex installation
- Higher initial cost
- May require more maintenance

Best for: Users wanting both dry and steam sauna options in one unit

Power range: 6-12 kW

Key features to consider:
- Steam generation capacity
- Ease of switching between modes
- Water tank capacity and accessibility

Comparison Factors:

1. Heat-up Time:
   - Infrared: Fastest
   - Electric: Quick
   - Gas: Moderate
   - Wood-burning: Slowest

2. Operating Costs:
   - Wood-burning: Potentially lowest, depends on wood availability
   - Gas: Often economical, varies by region
   - Electric: Varies based on electricity rates
   - Infrared: Generally energy-efficient

3. Maintenance:
   - Electric & Infrared: Low maintenance
   - Gas: Moderate (annual inspections recommended)
   - Wood-burning: Highest (regular cleaning and wood management)

4. Installation Complexity:
   - Electric & Infrared: Relatively simple
   - Gas: Requires professional installation
   - Wood-burning: Needs proper ventilation and chimney installation

5. Temperature Range:
- Wood-burning & Electric: Highest (up to 200°F/93°C)
- Gas: High
- Infrared: Lower (typically 120-150°F/49-66°C)

When choosing a sauna heater, consider your space, local regulations, available utilities, desired sauna experience, and long-term costs. Always prioritize safety and follow manufacturer guidelines for installation and use.

Glossary of Sauna Terms

1. Aufguss: A German sauna ritual where water, often infused with essential oils, is poured over hot stones to create steam bursts.

2. Bannik: In Slavic folklore, the spirit or guardian of the sauna.

3. Banya: A Russian-style sauna, typically wood-fired with high humidity.

4. Birch Vihta/Vasta: A bundle of birch twigs used in Finnish saunas for gentle self-flagellation to improve circulation.

5. Dry Sauna: A traditional sauna with low humidity, typically 10-20%.

6. Emitter: In infrared saunas, the device that produces infrared heat.

7. Far-Infrared: A type of infrared radiation used in some saunas, known for its deep tissue penetration.

8. Hammam: A Turkish-style steam bath, similar to a wet sauna.

9. Heat Therapy: The use of high temperatures for health benefits, as in saunas.

10. Hyperthermic Conditioning: The process of acclimating the body to heat stress, often through sauna use.

11. Infrared Sauna: A type of sauna that uses infrared heaters to emit radiant heat absorbed directly by the body.

12. Kiuas: The Finnish term for a sauna stove or heater.

13. Löyly: In Finnish sauna tradition, the steam that rises when water is thrown on hot stones.

14. Outdoor Sauna: A standalone sauna structure built outside, often near lakes or in natural settings.

15. Pefletti: A small, disposable seat cover used in public saunas for hygiene.

16. Sauna Hat: A wool or felt hat worn to protect the head from excessive heat in the sauna.

17. Sauna Master: An experienced person who leads sauna sessions, especially in public or spa settings.

18. Savusauna: A traditional Finnish smoke sauna, where the sauna is heated by wood fire without a chimney.

19. Steam Generator: A device that produces steam for wet saunas or steam rooms.

20. Steam Room: A high-humidity environment similar to a wet sauna, typically with 100% humidity.

21. Thermotherapy: The therapeutic use of heat, including sauna sessions.

22. Tylö: A Swedish sauna and steam bath manufacturer, often used generically to refer to modern sauna equipment.

23. Wet Sauna: A sauna with high humidity, created by pouring water over heated stones.

24. Wood-Burning Sauna: A sauna heated by a wood-fired stove, traditional in many cultures.

25. Barrel Sauna: A compact, cylindrical sauna shaped like a barrel, often used outdoors.

26. Contrast Therapy: The practice of alternating between hot (sauna) and cold (ice bath or cold shower) exposures.

27. Electric Sauna Heater: A sauna stove powered by electricity, common in modern home saunas.

28. Foot Bath: A small pool of cold water near the sauna entrance, used for cooling and cleansing feet.

29. Infrared Emitter: The heating element in an infrared sauna that produces infrared radiation.

30. Laude: Sauna benches, typically made of wood that doesn't overheat or splinter.

31. Mobile Sauna: A portable sauna built on a trailer or in a vehicle, allowing for sauna experiences in various locations.

32. Sauna Rocks: Special stones used in sauna stoves to store and radiate heat, and to create steam when water is poured over them.

33. Sauna Round: A complete sauna session, including heating, cooling, and rest periods.

34. Sweat Lodge: A ceremonial sauna-like structure used by some Native American cultures.

35. Thermometer/Hygrometer: Devices used to measure temperature and humidity in a sauna.

This glossary covers a wide range of terms related to various sauna traditions, equipment, and practices. Understanding these terms can enhance your sauna knowledge and experience.

Resources for Sauna Enthusiasts

As a sauna enthusiast, having access to quality resources can greatly enhance your experience and knowledge. Here's a comprehensive guide to various resources available:

1. Books and Publications

a) Essential Reading:
   - "The Sauna" by Rob Roy
   - "The Art of Sauna Building" by Bert Järvinen
   - "Sauna: The Finnish Bath" by H.J. Viherjuuri

b) Health-focused Books:
   - "Sauna Therapy" by Dr. Lawrence Wilson
   - "The Healing Power of Infrared Saunas" by Jennifer Egbert

c) Magazines and Journals:
   - "Sauna Journal" (International Sauna Association)
   - "Saunologian" (Finnish sauna blog and online magazine)

2. Online Resources

a) Websites:
   - SaunaScape.com: Comprehensive sauna information and reviews
   - SaunaVille.com: Community forum for sauna enthusiasts
   - TheSaunaPlace.com: Educational resources and product information

b) Forums and Communities:
   - Reddit r/Sauna: Active community for sauna discussions
   - Facebook Groups: "Sauna Enthusiasts Worldwide," "Sauna Building and Design"

c) YouTube Channels:
- "Saunatimes" - DIY sauna building and culture
- "Finnmark Sauna" - Traditional sauna practices and education

3. Associations and Organizations

a) International Sauna Association (ISA):
- Promotes sauna culture worldwide
- Organizes conferences and events

b) North American Sauna Society (NASS):
- Focuses on sauna traditions in North America
- Offers certification programs for sauna operators

c) Finnish Sauna Society:
- Preserves and promotes Finnish sauna culture
- Provides research and educational materials

4. Apps and Digital Tools

a) Sauna Timer Apps:
- "Sauna Workout" - Customizable sauna session timer
- "Sauna Tracker" - Log and analyze your sauna sessions

b) Temperature and Humidity Monitors:
- "SaunaLog" - Bluetooth-enabled sauna condition monitoring

5. Courses and Workshops

a) Online Courses:
- "Sauna Design and Construction" on platforms like Udemy
- "Traditional Sauna Techniques" offered by various cultural institutions

b) In-Person Workshops:
- Sauna building workshops (check local sauna suppliers or enthusiast groups)
- Sauna therapy certification courses

6. Suppliers and Manufacturers

a) Sauna Equipment:
- TyloHelo (www.tylohelo.com)
- Harvia (www.harvia.com)
- Finnleo (www.finnleo.com)

b) DIY Sauna Kits:
- Cedarbrook Sauna (www.cedarbrooksauna.com)
- Almost Heaven Saunas (www.almostheaven.com)

c) Specialty Supplies:
- Sauna Place (www.saunaplace.com) for accessories and parts

7. Travel and Experience

a) Sauna Tours:
- Finnish Sauna Tours (various operators in Finland)
- Banya experiences in Russia

b) Sauna Events:
- European Sauna Marathon in Estonia
- World Sauna Championships (currently discontinued, but similar events exist)

8. Research and Academic Resources

a) Scientific Journals:
- "Complementary Therapies in Medicine"
- "Journal of Human Kinetics" (for sauna physiology studies)

b) University Research Centers:
- University of Eastern Finland (conducts sauna health research)

9. Podcasts

a) "Sauna Talk" by Glenn Auerbach
b) "Sauna Podcast" by Lassi A. Liikkanen

10. Social Media

a) Instagram: Follow hashtags like #saunalife, #saunatime, #saunaculture
b) Pinterest: Great for sauna design ideas and DIY projects

Utilizing these resources can deepen your understanding of sauna culture, help you make informed decisions about sauna use or construction, and connect you with a global community of sauna enthusiasts. Whether you're interested in the health benefits, cultural aspects, or technical details of sauna building, these resources offer a wealth of information to explore.

www.ingramcontent.com/pod-product-compliance
Lightning Source LLC
Chambersburg PA
CBHW071914210526
45479CB00002B/418